Access Denied

Poverty, No End in Sight

by

Barthelemy Charles

PURPLE OWL PUBLISHING

Newton, MA

Table of Contents

Introduction

The more you know, the more you hurt; the more you understand, the more you suffer.
Ecclesiastes 1:18 (Contemporary English Version of the Bible)

In 2001, I was accepted into a two-year program of study in Sustainable International Development. In addition to the partial scholarship awarded by the university, I was drawn to the program because of the strong focus on poverty and inequality. It was primarily through the program that I became aware of my unilateral and myopic view of the causes and effects of poverty. For this reason, I am indebted to those who made my experience and awareness possible.

I benefited tremendously from interactions with classmates personally affected by civil war, genocide, and harmful cultural practices. Their experiences had been frightening yet edifying. My brain became flooded with questions, which propelled my desire to undertake this work. I wanted to understand how growing up in an impoverished community impacted my view of poverty and the consequences of my actions and inactions. To this end, I felt it necessary to explore the impact of decisions and actions at the household, community, national, and international levels on the plight of people living in poverty. In this book, I explore these actions through the lens of ideology, market forces, politics, the justice system, education, and religion, as well as select events that have occurred and continue to occur in St. Lucia, my place of birth. Given the relationship between violence against women and poverty, the book also discusses the subject and claims that there must be more substantive interventions and fewer symbolic gestures of change. The book further discusses the status of the elderly, considering the toll that poverty can have during old age.

For readers unfamiliar with Saint Lucia (St. Lucia), it is an island in the Caribbean with an area of about 238 square miles. The country had a population of 178,696 in 2018. Over the years, the island's economy has shifted from a reliance on the agricultural sector (mainly bananas) to the service sector (mainly tourism). It is the birthplace of two Nobel laureates, one in economics and the other in literature. A popular landmark, the Pitons, was designated a World Heritage Site by the United Nations Educational, Scientific and Cultural Organization (UNESCO). It has become a popular tourist destination and has received many prestigious awards. Many popular athletes, musicians, and television personalities frequent the island. But like any other country, St. Lucia has its own share of problems. In addition to a high level of unemployment at the time of writing, the country is grappling with a high incidence of crime and type 2 diabetes.

The proliferation of poverty in the richest and poorest countries presents daily opportunities to evaluate our personal and shared priorities. And while I will argue that poverty is woven into the fabric of our society, there are numerous opportunities for positive behavior to reduce poverty and mitigate its effects. When we look at the realities of poverty and allow our positive behaviors to be guided by self-sacrifice instead of self-interest, we can finally begin to improve conditions for those who have been denied access to a better way of life.

Chapter 1:
Reality Check: Access Denied

"The destruction of the poor is their poverty."
Proverbs 10:15 (King James Version of the Bible)

The renowned reggae artist, Bob Marley, popularized the biblical quote mentioned in one of his songs, Wisdom. True to these words, the poor have been locked out of opportunities for social and economic development by virtue of their position in society. Sadly, throughout history, humans have experienced and recovered from several deadly and indiscriminate epidemics. Diseases such as the bubonic plague ended the lives of millions of people in Europe between 1347 and 1351 (Encyclopedia Britannica, 2023). Before its eradication in 1977, at least 300 million people died of smallpox in the twentieth century (National Geographic, 2023). And as of 6:33 p.m. (CST) on September 2, 2022, the World Health Organization (WHO) reported 6,475,346 confirmed deaths from the coronavirus pandemic. However, the negative impact of the bubonic plague, smallpox, and other deadly diseases pales in significance when compared to the cumulative number of lives affected by poverty. This global problem, which denies its victims access to basic needs, especially food, clothing, and shelter, merits greater attention since it threatens the lives of billions of people worldwide.

Extreme Poverty

In 2000, the decision by 189 member countries of the United Nations to adopt the Millennium Declaration, and by extension, the Millennium Development Goals (MDGs), proved that at the highest political level people were aware of the magnitude of global poverty and the dangers it poses. A popular measure used to determine the extent of global poverty is the international poverty line devised by the

World Bank. It measures the number of people living in extreme poverty and allows for comparisons between countries. In 1993 the international poverty line was US$1 per day. In 2005 it was recalibrated to US$1.25, and again in 2015 to US$1.90 per day (World Bank, 2015). One of the many goals of the MDGs was to reduce the number of people living on those low amounts by 2015.

The 2009 Millennium Development Goals (MDGs) Report published by the United Nations highlighted that from 1990 to 2005, the number of people living on less than US$1.25 decreased from 1.8 billion to 1.4 billion (p. 4). In 2008, about 110 million fewer people than the 1.4 billion recorded in 2005 lived in extreme poverty (United Nations [UN], 2012, p. 7). The year 2015 marked the end of the Millennium Development Goals (MDGs), and the report for that year indicated that the number of people living on less than US$1.25 per day was 836 million (p. 4). Of note, given limitations in areas such as data capture, quality, and monitoring in some reporting countries, it was reported that the poorest people in these countries often remain invisible. If we take this information into account, the number of people living in poverty is far more than reported.

Beyond the Poverty Line

To highlight the dangers of poverty, some key findings in the previously mentioned MDGs report for 2015 are worthy of note:

- Poverty is a major determinant of malnutrition, which continues to affect millions of people worldwide. The number of undernourished people worldwide was an estimated 795 million (p. 20), with the majority residing in developing countries.

- Besides malnutrition, inadequate water supply and sanitation are linked to poverty. While the goal set for access to safe drinking water had been met, 663 million people still had unimproved access to water, and 2.4 billion people still used unimproved sanitation facilities (p. 58).

- Inadequate housing is another growing problem facing millions of people living in poverty. An estimated 880 million urban residents lived in slum conditions, and some increases were because of the accelerated pace of urbanization (p. 60).

- People living in poverty are the most likely to suffer from sexually transmitted and other diseases due to inadequate health education and limited access to quality health care services. In 2013 there was an estimated 11 million people living with tuberculosis (p. 49) and an estimated 35 million still living with the human immunodeficiency virus (p. 45). An estimated 17.7 million children under the age of eighteen lost one or both parents to AIDS-related causes in that same year (p. 46). In 2015, there were an estimated 214 million cases of malaria (p. 47).

- Disadvantaged people are often the least educated and thus less likely to be employed. An estimated 57 percent of primary-school-aged children were out of school in 2015, and it was anticipated that, based on 2012 estimates, 43 percent of school-aged children will never attend school (p. 25). In some countries it was also found that children from poor families are four times less likely to attend school than those from richer families (p. 24).

- People living in poverty are often the most affected by natural disasters and conflicts. Hence, "by the end of 2014, conflicts had forced almost 60 million people to abandon their homes" (p. 8).

It is important to note that some displaced people are driven into poverty due to a lack of employment opportunities, among other things, in host countries with limited resources and poor accommodations. While it is hoped that these displaced people would be welcome in their host countries, migrants may be unwelcome due to the rise of nationalism in some parts of the world—for example, in some parts of Europe (Trilling, 2019).

Regarding educational outcomes, recent estimates (Walker et al., 2019, p. 4) suggested that "across the developing world, a child from a poor family is seven times less likely to finish secondary school than a child from a rich family."

The final report of the MDGs in 2015 also highlighted the following achievements: The number of people living in extreme poverty reduced by more than half over the period 1990–2015 (p. 4); 2.6 billion people had access to an improved drinking water source (p. 7); the number of primary-school-aged children out of school fell by almost half from 100 million in 2000 to an estimated 57 million in 2015 (p. 4); antiretroviral therapy saved the lives of 7.6 million people with AIDS over the period 1995–2013 (p. 6); the incidence and mortality rates in malaria and tuberculosis declined; and child (under five) mortality declined by more than half (p. 6). It is, however, uncertain how sustainable these achievements will be, as history has shown how they can erode overnight.

The Sustainable Development Goals (SDGs)

The end of the Millennium Development Goals (MDGs) marked the beginning of the Sustainable Development Goals (SDGs) and an international poverty line of US$1.90. A new target was established, namely to eradicate extreme poverty for all people everywhere by 2030. The SDGs are supposed to build on the achievements of the MDGs, requiring all countries, including poor countries, to promote prosperity while protecting the planet. The challenges that lie ahead were captured in the words of the then Secretary General of the United Nations (2013): "Achieving the 2030 Agenda requires immediate and accelerated actions by countries along with collaborative partnerships among governments and stakeholders at all levels." Notable events after the close of the MDGs, such as the coronavirus pandemic and the war between Russia and Ukraine, further highlight the daunting task ahead. The SDGs report for 2021 showed that 119 to 124 million people were pushed back into extreme poverty because of the COVID-19 pandemic (p. 2).

Points to Ponder

The preceding statistical data provides a glimpse into the number of people being denied access to basic resources for survival. The importance of these basic needs transcends cultural, social, and religious boundaries. While some choose to ignore the fact that countless people are denied access to basic needs, the consequences do not go away. When people are desperate, they take drastic and desperate measures. Hence, some countries have experienced a high influx of refugees as people take great risks on land and sea to escape poverty and events like wars, earthquakes, and other natural disasters that can make people destitute within minutes. As the number of refugees increases, previously accommodating governments and citizens begin to realize the magnitude and implications of the crisis; foreigners, whether newly arrived or established, may no longer be welcomed.

When people are denied access to life's necessities, some in their hopelessness develop a hatred for whatever reminds them of the status quo and become more open to ideologies and movements that provide a sense of being and purpose. The aim of some of these ideologies and movements is to change the balance of power and ownership of natural resources.

Historically, any movement to change the balance of power and ownership of resources has human casualties, including those who sacrifice their lives for the cause and bystanders. Really, then, ensuring that people have access to life necessities is in our best interest. Some respond to this by supporting charitable organizations. Others lobby their governments to create enabling environments that provide opportunities for people to improve their lives. Many put their lives at great risk to be physically present in the delivery of goods and services to help the needy build their capacity. However, it remains to be seen whether our collective efforts will be sufficient to withstand the collective force from those who benefit from the status quo.

Moreover, once we continue to see each other as different based on skin color, there will always be prejudice and discrimination regarding who is entitled to natural resources and social goods. Once we cherish our national identity more than the bond we share as humans, there will always be territorial conflicts uprooting the lives of people and leaving them destitute. Regrettably, while the debate on the best way to distribute these basic requirements continues, this fact remains: the world's poor are denied access to the basic needs they are entitled to by virtue of their existence. Citizens of the earth should have mutual ownership of its resources.

Chapter 2:
Ideology Versus Reality

"Sustainable development is the pathway to the future we want for all."
United Nations Secretary General (2013)

Over the years there has been a range of theories and ideologies that have emerged with the common objective of addressing the plight of the disadvantaged. With support from governments and other influential institutions, many of these ideas have materialized into realities, some successful and others detrimental. In recent years, the ideology of sustainable development has gained prominence, championed by many world leaders and influential institutions. A closer look at past and unfolding debates and controversies regarding sustainable development reveals a daunting task ahead for proponents of this altruistic ideology.

Polarization of Will and Power

At first glance, the ideals of sustainable development seem reasonably attainable. In its simplest terms, sustainable development is "development that meets the needs of the present without compromising the ability of future generations to meet their own needs." This would translate to exploiting natural resources in harmony with the earth's regenerative capacity and the fair distribution of these resources. However, quantifying the earth's regenerative capacity and determining the needs of current and future generations is, by nature, complex. This has fueled the disagreement between politicians and scientists on addressing issues pertinent to sustainable development, such as global warming. Political differences exacerbate this problem and serve to either stagnate decision-making or defer commitment to a resolution.

Armed with the experience and knowledge of previous attempts, the tenets of sustainable development are conceptualized, at least in theory, to engender a new genre of policies centered on an integrated approach to address global poverty. In principle, this integrated approach allows beneficiaries to be actively involved in planning and implementing related policies and interventions.

In this all-encompassing approach, the actors responsible for this transformation can be grouped into three broad and sometimes overlapping levels: at the low level, the poor and disadvantaged; at the middle level, practitioners and theorists; and at the high level, world leaders and powerful financial and political institutions. Consumers (because of their purchasing power) and advocacy groups (because of their strategic alliances) are distributed at all levels, and they exert much power and influence on the matter. However, consumers' decisions are usually more a matter of price as opposed to equality and justice, and some advocacy groups are sometimes stifled by their funders.

At the low level, the poor and disadvantaged are generally open to ideas, theories, or social movements that can elevate their social status. They are highly committed to new ideas that affect them. However, their commitment may be a consequence of their existing socioeconomic status and is likely to change as their status improves. Accounts of successful projects involving microfinance and integrated conservation are indicative of the commitment of disadvantaged people to sustainable development. Their openness to being educated and their willingness to discontinue cultural practices that counter development have yielded some favorable results. Still, the overall outlook for many disadvantaged people remains bleak as they lack the resources to bring about desired changes. By virtue of their circumstances, they often contribute to and exacerbate the problems they wish to eradicate. They are often faced with disadvantageous and limited choices—for example, having to choose between curbing their hunger or the further exploitation of already depleted natural resources.

At the middle level, theorists and practitioners generally seek to transform academic and political solutions into tangible realities. They are a crucial link between the powerless and the powerful. Opportunities for available funding and personal or institutional recognition sometimes drive their agendas. Hence, they may, at times, advance the agenda of funding and research agencies that are not always in the best interest of stakeholders at the low level. Moreover, it would take a great deal of humility to allow the disadvantaged to steer the wheel of development and let them take over.

At the highest level, decisions made on sustainable development are almost always tied to economic and political advancement. Some powerful world leaders vacillate on their commitment to addressing issues threatening sustainable development such as global warming. The scientific arguments and evidence of global warming and the projected negative impact on both the poor and rich are becoming too buoyant to ignore. This has sparked powerful decision-makers to act or, at least, commit to act. However, some of these actions are driven by opportunities to generate profits and maintain the status quo, while others are very elaborate with no specifics or just simply unattainable (at least for now).

Therefore, it is highly unlikely that concrete results will follow soon, as there are already foreseeable obstacles at the high level—namely, the battle for the ownership of the resources in the Arctic, which may become accessible due to global warming (Arsenault, 2012). This is a potential disincentive to tackle global warming, and people living in poverty will continue to suffer while others plan to reap the "benefits" of global warming.

Given the diversity of actors and the polarization of will and power, the ideals of sustainable development may fall short of addressing the plight of people living in poverty in the way it proposes.

A Preoccupation with Economic Growth

As Esty (2001) has argued, while the conservation movement of the 1970s was concerned with the preservation of the earth's resources, sustainable development tries to marry economic development and environmental preservation. This marriage is an attempt to shift environmentalist focus away from the proposed view that the current rate of economic growth is incompatible with environmental conservation. Even so, economic growth generally continues to dominate decision-making at the individual, family, and national levels. The social and environmental costs may alter decisions if they pose an immediate threat to growth. The fast economic growth in some developed countries—at least before the 2008 global financial crisis—did not come about without tremendous damage to the environment and the health and safety of certain sectors of the population.

This preoccupation with economic growth at the macro level is partly sustained by the notion that the benefits of such growth will filter down to the poor. Despite this notion, many people continue to lose their jobs daily as businesses shut down, outsource, relocate, restructure, merge, and automate their operations to maximize shareholder profits or just stay in business. Stakeholders at the low level, lacking in skills and education, are the most vulnerable to the negative impact of these practices. Obviously, once people cannot afford to take care of their personal responsibilities—their basic needs—we cannot expect them to take care of shared responsibilities. For example, there is concern over the loss of biodiversity due to deforestation in countries like Haiti (Hedges, Cohen & Yang, 2018). This is most likely due to land clearing for agriculture and wood for charcoal production. Of note, Haiti is one of the poorest countries in the Caribbean, and environmental preservation may not be a pressing matter on the minds of people living in poverty and widespread civil unrest. Nevertheless, finding the right balance, if possible, between economic growth and human development will be critical to achieving real success in such matters as environmental preservation.

Development Aid, Budget Allocation, and Corruption

Development assistance from the most influential and financial global players is important to impoverished and developing countries. Some more developed nations have pledged 0.7 percent of their gross national income (GNI) as official development assistance (ODA). The MDGs report of 2012 noted that "in 2011, net aid disbursements amounted to $133.5 billion, representing 0.31 percent of developed countries' combined national income" (p. 58). The 2014 report highlighted that $134.8 billion was disbursed in 2013, which represented 0.3 percent of the combined gross national income of developed countries (p. 48). The last MDGs report of 2015 noted that official development assistance from member countries of the Development Assistance Committee of the Organization for Economic Cooperation and Development (OECD) represented ".29 percent of their gross national income in 2014" and that some member countries, such as Denmark, Luxembourg, Norway, and the United Kingdom, exceeded this target of 0.7 percent of gross national income (p. 62). In the SDGs report for 2018, it was noted that "net ODA totaled $146.6 billion and ODA as a share of donors' gross national income (GNI) remained low, at 0.31 percent" (p. 13). The Inter-agency Task Force on Financing for Development of the United Nations (2021) reported that "ODA increased by 0.7 percent in 2019, to $155 billion in real terms" (p. 82). However, this was still well below the committed 0.7 percent of gross national income.

While it is clear that there is a shortfall in official development assistance, the question remains whether assistance from other sources could make up for the shortfall and whether this level of ODA (0.7 percent of GNI) is sufficient to achieve sustainable development targets for impoverished countries. Moreover, it is important to ascertain the percentage of ODA that is not used according to plan and the degree to which it impedes success. The presence of corruption in the public sector is not new, as highlighted by the anticorruption watchdog, Transparency International (https://www.transparency.org). Of the

180 countries and territories assessed in 2018, two-thirds recorded a score of less than fifty on the Corruption Perceptions Index (CPI), which uses a scale of 0 to 100, where 0 is highly corrupt and 100 is very clean. Among the areas covered by the CPI are the rates of bribery, diversion of public funds, and conflicts of interest. It is very interesting to note that no country received a score of 100, which suggests no country is corruption-free. Therefore, any assistance to poor and developing countries must factor in the extent of public sector corruption to determine the real value of such assistance.

On a national level, the allocation of resources between various government sectors is also an indicator of the commitment to address the plight of the disadvantaged and the commitment to development that is sustainable. It is often said that decisions are taken within the context of resource scarcity and are expected, at least in principle, to benefit all citizens. However, the burden is often disproportionately placed on disadvantaged people, as programs that serve their needs are usually affected in times of resource constraints that seem endemic to poor countries. These programs are sometimes terminated, underfunded, and poorly supervised.

The level of spending and commitment that may address the plight of the poor is often seen in the wake of disasters, natural or man-made. But the outburst of tears, the inflow of funds, and the delivery of emotional speeches can never make up for the number of lives lost and affected by such disasters, including the consequential economic, emotional, and psychological damage. The influx of funds from private and public sources often gives rise to mismanagement and corruption in these poorer areas. Conveniently, the impulse to donate to countries impacted by disasters often fades over time, and this helps to maintain the status quo.

Points to Ponder

Just as generations come and go, so do economic and political ideologies. These ideologies have at least two common objectives: to protect global resources and determine the means of distribution among competing parties. Many ideologies seem ideal on paper but fall short in reality, and it is usually the ideology with adequate financial, political, and military support that goes through the experimental phase and beyond. The wisdom behind any ideology is proven by its outcomes.

There is great financial and political support behind sustainable development. However, while there has been significant progress in key areas, it remains to be seen whether there is sufficient support to reach the desired goals. To many, it seems irrational to be overly concerned with the needs of future generations when their immediate needs cannot be satisfied. Others continue to destroy the lives of their fellow men and the environment on the premise that the consequences will not likely take effect during their own lifetime. Some deliberately withhold the means of survival from others for their own political and selfish gains. Moreover, the developmental goals in themselves may need to be more attainable. In addition, the ideas that could bring about global, stable changes will fail to transcend unless those at the steering wheel of the political and financial institutions can experience the inadequacies of public services and institutions and confine themselves to the same working and living conditions as the poor. This, however, does not shield the rest of us from personal responsibility.

Several years ago, I strongly believed that a good deed in a remote community would not create an impact if people in the more affluent countries did not follow suit. However, I have come to realize this kind of mindset creates and maintains the conditions of poverty and inequality around us. It does not help to justify our inaction by pointing to the faults of others. I have also come to acknowledge that the popularity and wide acceptance of a course of action do not necessarily mean that it is ideal or appropriate. More importantly, it is becoming clearer to me that it would require a force above us and a cause beyond us to bring about and sustain the universal changes that would benefit all, including the poor and disadvantaged.

Chapter 3:
St. Lucia and the Sustainable Development Goals

It is better not to make a vow than to make one and not fulfill it.
Ecclesiastes 5:5 (New International Version of the Bible)

In 2015, St. Lucia committed to the 2030 Agenda for Sustainable Development. St. Lucia has since established a governing committee to guide the implementation and monitoring of the Sustainable Development Goals (SDGs) locally. There has been some level of stakeholder interaction around prioritizing the key issues with associated goals and targets. However, some key challenges have been identified, including human, financial, technological, and institutional capacity constraints. To highlight what would be required to translate the SDGs into realities, we will examine some limitations that may impede or delay successful outcomes for St. Lucia.

SDG 2: End hunger, achieve food security and improved nutrition, and promote sustainable agriculture.

Traditionally, poor people in St. Lucia with access to land, whether through inheritance or squatting, produce most of what they need for consumption, and surpluses are offered up for sale to generate revenue to purchase other essential supplies. Lately, however, the food security of many in the rural communities of St. Lucia has been threatened by the growing prevalence of predial larceny, the theft of agricultural crops. An issue brief by the Food and Agriculture Organization (FAO) of the United Nations (2013) on predial larceny in the wider region, the Caribbean, noted that it is the most extensive among all crimes committed based on the number of individuals and families affected.

Additionally, "more than 90% of producers regionally agree that it is the single greatest disincentive to investment in the sector" (p. 1).

It is believed that many individuals suffering from drug and substance abuse are responsible for these acts, selling stolen goods to purchase illicit drugs. However, in most cases, these stolen goods are allegedly sold to local vendors/hucksters at a minimum cost. Sadly, sometimes these crops are treated with pesticides that render them unfit for consumption many weeks after application. Nevertheless, they find themselves in the homes of consumers days after treatment application.

The government of St. Lucia established a patrol unit to address the situation, and the public is encouraged to report incidents of theft. Based on data from the Central Statistical Office of St. Lucia (https://stats.gov.lc), the overall number of cases of predial larceny has since declined. However, the number of arrests and convictions is relatively low compared to the number of cases that are reported and substantiated. Of note, the FAO (2013) indicated that only 45 percent of incidents are reported to the police regionally due to frustration with inaction by authorities.

Nevertheless, many people suffering from drug and substance abuse, and others, have since found themselves in prison—some repeatedly. It could be argued that this jailing response does not discourage this practice as long as markets are willing to purchase these stolen goods. Therefore, it can be argued that greater efforts should be made to apprehend and convict people who knowingly buy these stolen goods, both household members and vendors. This, however, is not as easy as it may sound, given that some of the stolen items "are used for household food and other needs in the home" (p. 3), based on data from the wider region published by the FAO (2013). Hence, the solution has moral implications, as there may be a much deeper reason rooted in poverty that some would argue justifies the theft that puts food on the table for these poorer families. Of note, it was reported that the number of cases in St. Lucia increased during the

COVID-19 pandemic (Antione, 2021), possibly due to the high food prices and unstable supply chains.

Until a proper and sustained mechanism can be put in place to deal with this problem, some people will have difficulty producing the foods they need. This is unfavorable in an economic climate where the overall cost of goods and services escalates. The coronavirus pandemic has highlighted the susceptibility of global supply chains. The predicted negative impact of climate change, such as prolonged droughts and crop failures, signals the possibility of countries restricting exports if it threatens local food security. This all points to the importance of local production to food security. Therefore, the government of St. Lucia must ensure that it addresses all hurdles that could discourage local crop production, including the theft of agricultural produce.

SDG 3: Ensure healthy lives and promote well-being for all at all ages.

People living in poverty generally spend their daily income on basic needs for survival and sometimes cannot afford healthy food options that could lead to better health outcomes. There is often too little income left to save, let alone to address emergency situations or plan for inevitable problems, especially those associated with aging. Based on a small study (Holden et al., 2015), it was found that unemployment, low income, limited secular education, and poor insurance coverage were determinants of healthcare access in St. Lucia (p. 10). As of 2019, heart disease, cancer, stroke, and diabetes were among the leading causes of death on the island. Some of these conditions are linked to a poor diet and physical inactivity.

A considerable share of the population, especially the poor, relies on the public health care system for affordable and subsidized service since only a small share of the population has private insurance coverage, a mere ten percent in 2014 (World Bank, 2018). This is consistent with a technical report on health care financing in St. Lucia and costs of Victoria Hospital (Russell et al., 1988) conducted as far back as 1988 under the auspices of the United States Agency for International Development (USAID). The report indicated that "there is a large low-income group for whom substantial out-of-pocket expenditures for health are not possible" (p. iii). In 2018, the World Bank highlighted that out-of-pocket payment for health care services in St. Lucia was almost 46 percent, over twice the target recommended by the World Health Organization (p. 26). This is of serious concern since a high number of St. Lucians do not have private health insurance. The World Bank further cited St. Lucia's Ministry of Health and Wellness (2018) as it revealed that less than ten percent of the population had private health insurance and that coverage was concentrated among the higher income quintiles (p. 5).

However, the public health care system in St. Lucia is limited in several ways. Often, key drugs and services are not available and have to be purchased from private providers, often making them out of reach for disadvantaged people. While some disadvantaged people qualify for free or discounted services, these are sometimes unavailable or inefficient due to systemic breakdowns, including inadequate financial, technical, and infrastructural support.

On the night of September 8th, 2009, sections of St. Jude Hospital, one of the major hospitals on the island, were destroyed by a fire that tragically ended the lives of three patients. The fire incident caused the hospital to move operations to the nearby national stadium. Since then, some rebuilding work has been undertaken, and the hospital was still incomplete at the time of writing.

The financial contributions by the government and payments from patients were expected to cover the cost of operations at the hospital. However, a 2004 report by a team assembled to review the terms and conditions of employment of medical practitioners and nurses noted that the hospital had persistent difficulties "in collecting fees from Government agencies and/or persons receiving services on behalf of Government such as prisoners, nurses, police officers and firemen"

(Haynes et al., 2004, p. 84). Of note, as far back as 1988, the previously cited USAID report highlighted that it was unclear that the annual subvention that the national insurance program (NIS) contributed to the Ministry of Health was "either appropriate under current legislation or adequate for coverage of health care costs incurred by NIS enrollees" (p. ii). A later assessment of the government expenditure on health (Rodriquez et al., 2012, p. 29) hinted that the government should allocate more funds to healthcare based on a comparison with other Caribbean countries. In addition, it was subsequently reported that public sector health expenditure was still below the threshold recommended by the World Health Organization (World Bank, 2018, p. 7).

Payments from the public were also crucial for the effective functioning of St. Jude Hospital. However, the public was delinquent in the payments for services rendered by the hospital. Some people genuinely could not afford to pay, even when those services were subsidized. With the fragmentation of health care services, people are often reluctant to make a payment to a public hospital when only part of their required services are available there. Instead, their funds go to services obtained from private providers.

The Victoria Hospital (VH), the largest hospital up until 2020, has been under constant criticism regarding staffing and the extent, quality, and sustainability of services. The following are just a few of the complaints cited in the report of the 2004 Health Review Commission (p. 52) against some doctors in the public health service:

- unauthorized removal of medical equipment from VH to use elsewhere.
- priority given to patients admitted to the hospital from private offices.
- inappropriate influence on patients to use private laboratories in which doctors have a financial interest.

- out-patient clinics often left to junior doctors to manage while senior staff are in their private offices.
- ward rounds conducted by telephone and other instances of hurried rounds by doctors eager to leave for their private offices.
- collusion with some nurses to obtain priority treatment for private patients.
- unavailability/inability of doctors to respond to urgent calls because of commitments to private practice, even when they are "on call" for emergencies at the hospital.

While these observations are significant, some may be triggered by the limitation of services in public facilities and the high cost in private facilities, pushing both medical staff and patients to find unconventional ways to provide and receive medical care. The report said, "There is an abundance of evidence which allows us to conclude that some doctors employed in the public service devote as little time as possible to their public duties and seek to maximize their earnings in the private sector. This is an issue which concerns most Saint Lucians, especially those who cannot access private medical care and are subjected to what they consider as sub-standard care in the public sector" (p. 52). Thankfully, in 2018, funding in the amount of US$20 million from the World Bank was secured for the implementation of the Health System Strengthening Project for St. Lucia to improve the accessibility, efficiency, and responsiveness of key health services.

In March 2020, a new national hospital, the Owen King EU Hospital, became fully operational. However, the move to the new facility does not guarantee improvement in health outcomes unless the observations of the 2004 Health Review Commission, if substantiated, are addressed. In addition, the migration of healthcare professionals to countries that offer better financial incentives is also an area of concern. Nevertheless, the cost of health care services at public facilities in St. Lucia is relatively cheap, well below the actual cost—a point that cannot be overemphasized. There is a network of health centers around the island serving various communities in which qualified citizens receive medications and other treatments, sometimes at minimal or no cost. Nurses, doctors, and health aid workers attached to these health centers make numerous home visits at no direct cost to patients. It would make a great difference if patients would try their best to pay these basic costs and pay outstanding bills within a reasonable time frame. Citizens could also reduce the load on the health care system by practicing preventative care to the extent possible. However, many would agree that unless the policy and decision makers are confined to the same health care services that are meted out to the public, especially to people living in poverty, the health care system will continue to fall short.

SDG 8: Promote sustained, inclusive, and sustainable economic growth, and promote full and productive employment and decent work for all.

Is the goal of "full and productive employment and decent work for all" realistic? To economists, there is a vast difference between full employment and zero unemployment; the former is deemed good for the economy, and the latter is not so good. Theoretically, full employment means that unemployment has fallen to the lowest possible level, which won't cause inflation and keeps the market competitive. This suggests that to maintain a strong economy, some level of unemployment is beneficial.

Government officials have given themselves the mandate to create jobs, whether directly or through the creation and maintenance of the enabling environment. However, while governments endeavor to create jobs, it seems almost impossible to bridge the unemployment gap. In 2018, the labor force in St. Lucia was made up of 58.9 percent of the population. The overall unemployment rate was 20.2 percent, compared to a youth employment rate of 36.3 percent. The coronavirus pandemic worsened the poor job prospects for many due to the negative impact on the tourism industry, the major source of employment on the island. It is clear, then, that many people simply do not have jobs to provide the income to purchase the bundle of goods deemed necessary for decent living.

A government's responsibility does not stop with employment creation but must also encompass worker productivity and efficiency, as the lack thereof will negatively impact available resources. In fact, sustainable development demands the judicial use of resources. Regarding productivity and efficiency in the public sector, St. Lucia is not exempt from the challenges facing similar bureaucracies elsewhere that impede economic growth. Let us discuss at least three of these common observations. First, while service delivery is weak, "public sector wages and salaries usually take up a large share of public expenditure" (Schmid, 2018). In 2012, the government of St. Lucia indicated that in terms of recurrent expenditure, "nearly forty cents of every dollar" go toward wages and salaries. In 2022, it was noted that "from year to year, employee compensation represents on average approximately 33.25 percent of the annual total budget" (Government of St. Lucia, 2022).

Second, "strong civil-service rules and employee protections make it difficult to deal with poor performers" (Lavigna, 2014). Regrettably, some people generally view the public sector as a haven. Ironically, while many seek employment in the public sector because of the strong employee protection, others who view themselves as more progressive turn to the sector as a last resort. This often leads to a lack of commitment to whatever post they may occupy in the public sector. While some may gradually develop a fondness for the positions they hold, others, however, are only concerned with the bottom line—the salary at the end of the month. Still, others seem to enter public service to benefit from the scholarship programs and other incentives. Lately, the government has been hiring more workers on a contractual basis. This approach may make it much easier to dismiss poor performers, but without proper supervision, it can also be used to dismiss critics.

A third factor that contributes to the poor performance of the public sector is the way some of its employees are hired. There are "frequent and abrupt changes in leadership" and "many government jurisdictions and agencies are led by elected or politically appointed leaders (including legislators) with brief tenures, specific policy agendas, and short-term perspectives" (Lavigna, 2014). The frequent and abrupt change in leadership is evident in the projects established by the government to provide technical and sometimes financial support to banana farmers in St. Lucia. Sadly, the names, functions, and staff members on these projects have changed considerably with almost every new government. This has resulted in an incoherent approach to tackling the problems threatening the existence of the banana industry and the livelihood of local farmers. This has also prevented the industry from taking advantage of the knowledge and skills of some key individuals with decades of experience.

How do these three factors affect disadvantaged people and the greater goal of sustainable development? To begin with, services directed to people living in poverty will not always be managed or delivered effectively. The remark by Schmid (2018) on the constraints against growth and development in the Caribbean is significant: "Service delivery is also influenced by red tape as a result of a paper-based culture, regulatory complexity, insufficient coordination among ministries, and lack of digitization and use of information technology solutions."

Second, the government needs to allocate revenue to cover salaries for public employees, and since some are unproductive in certain areas, this means that the government is likely to incur losses as a result. This no doubt affects the government's financial/budgetary allocation directed towards alleviating poverty. Therefore, while the government may be the target of criticism for the status of citizens living in poverty, we must not forget that all those who man the bureaucracy play a part in creating the conditions that either alleviate or promote poverty.

SDG 11: Make cities and human settlements inclusive, safe, resilient, and sustainable.

To address the demand for decent and affordable housing, the government of St. Lucia has established several housing projects and made land available for sale, allowing some low-income citizens to purchase land that some of them have occupied for years as squatters. However, a major obstacle in obtaining real estate and decent housing, especially for disadvantaged people, is the ability to pay for real estate. This problem is now further compounded by the loss of employment and economic downturn due to the coronavirus pandemic.

It could be argued that most workers have monthly wages ranging from about EC$1,124 to $1,584 (US$420 to $593) or less. This is the salary range for frontline/operational staff in the hotel sector (Shahidsaless et al., 2018, p. 23), which can serve as a proxy for most people in the workforce since the hotel industry supplanted the banana industry as a main source of employment. Of note, overall wages may have declined with the advent of the COVID-19 pandemic due to possible greater competition for available positions.

Low-income earners are less likely to qualify for mortgages from financial institutions, given their high financial risk, which is further compounded by the new trend in both the public and private sectors to provide more short-term employment with no benefits. Further, the loans that low-income people do qualify for are usually short-term with higher interest rate payments.

Therefore, poor households are most likely experiencing housing affordability problems, both in terms of owner-occupied and rental housing (St. Lucia housing policy, 2008, p. 5). It is indeed a distressing situation when crown lands in a community are converted to housing projects or salable lots and low-income residents cannot purchase them. Such a situation will likely create tension among new occupants, usually from different communities.

Low-income families that can secure land, whether through purchase or inheritance, do not always build dwellings that meet the standards set by the local authorities. Moreover, some would argue that the money and time needed to obtain the necessary approval could be better spent on building materials and construction. Global warming is an impending threat for those living and still constructing structures along coastal seas, rivers, and ravines. The telling signs of what could occur should weather conditions change for the worse are constantly witnessed. With any sustained downpour, certain communities become isolated because of flooding.

In such times, people living in substandard dwellings with no property insurance sometimes risk their lives trying to safeguard their families and material belongings. Some of these incidents are reported in newspapers, in official government reports, on televised broadcasts, and on social media platforms. As previously mentioned, people in high-risk areas cannot usually afford the land or houses put up for sale on some private and government housing development projects. There must be some alternative to address their situation as the country strives to meet the sustainable development goal of making settlements inclusive, safe, resilient, and sustainable. In the interim, this would include the de-silting or clearing of rivers and other waterways, which has begun in some areas. Building bridges of appropriate height that will still be accessible after the expected heavy rains associated with the rainy season have been long delayed in well-known flood-prone areas, though some progress has been made to date. The long-run solution would include relocating entire communities, which has already occurred in some instances. However, an effective mechanism must be in place to deal with the expected resistance from community members who wish not to relocate for economic and nostalgic reasons.

Concessions on building materials for those whose income is below a justifiable threshold would also be of help. These concessions, however, must be closely monitored to ensure that they reach the target group. The simplification of submitting and obtaining approval for building plans would be a major feat. The timely identification and halting of construction of houses in disaster-prone areas, as well as greater availability of suitable and affordable land for purchase or rent within the shortest possible time frame, are equally important.

We must not forget that some people are just not able to afford suitable housing and that they would need full support. Governmental programs such as "Koudmein St. Lucie" have made it possible for some to gain decent housing. This particular program is designed to give support to families living in extreme poverty and is implemented by the St. Lucia Social Development Fund (www.slusdf.net). In 2021, the government further announced the launch of an EC$10 million National Housing Assistance Programme (St. Lucia Times News, 2021). It was reported that the program "will identify and create an inventory of households with the most urgent housing needs, assess their priority, design suitable assistance packages, and disburse them in the form of technical assistance, building material, and/or labour grants." Hopefully, this program will provide disadvantaged people with suitable housing. When people spend less time worrying about adequate shelter, they become less distracted and can focus their energies on more strategic and long-term activities that contribute towards sustainable development.

Points to Ponder

> "There was a small city with a few men in it; and a mighty king came against it and surrounded it and built great siege works against it. In it was found a poor but wise man, and he saved the city by his wisdom. But no one remembered that poor man."

The above is taken from an ancient book, and the writer is said to be the wisest man who ever lived. True to his words, people living in poverty have in some ways saved many "cities" to the extent that they unlock opportunities for financial and technical support from donor countries/agencies and money-lending institutions, leading to job creation and institutional/infrastructural improvements. Regrettably, sometimes people living in poverty are forgotten as the temptation to divert funds away from their needs to other competing priorities is great. The sustainable development goals have unlocked many opportunities, and it is hoped that the needs of people living in poverty, including those from St. Lucia, remain on the local agenda and are pursued with the same tenor as conveyed in the multiplicity of reports, meetings, and conferences.

Chapter 4: Market Forces: Unjust, Unethical, but Necessary?

Whatever affects one directly, affects all indirectly. I can never be what I ought to be until you are what you ought to be. This is the interrelated structure of reality.
Martin Luther King Jr.

Many scientists and other influential figures have voiced dire predictions for Earth and its inhabitants. Entrepreneur and business magnate Elon Musk noted that "either we spread Earth to other planets, or we risk going extinct" (Davies, 2013). Renowned physicist Stephen Hawking said, "Although the chance of a disaster to planet Earth in a given year may be quite low, it adds up over time, and becomes a near certainty in the next thousand or 10,000 years" (Rincon, 2018). In the dark days ahead, if the effects of global warming prove true, we will see more natural disasters, mass extinctions, accelerated depletion of the ozone layer, food shortages, and adverse weather patterns. But while global warming is given extensive media coverage and public attention, the problem of the vast income and wealth disparities worldwide is given less regard. The Global Wealth Report for 2022, published by the Credit Suisse Research Institute, estimated that "the bottom 50% of adults in the global wealth distribution together accounted for less than 1% of total global wealth at the end of 2021" whereas "the top richest decile (top 10% of adults) owns 82% of global wealth" (p. 32). It is hard to envisage a sustainable society with wealth inequalities where much of this wealth is generated through the workings of the market economy and the policies established by governments and other regulatory bodies worldwide.

While humans have utilized various forms of government to oversee them and trade systems to manage the seemingly scarce resources of the world, people have generally come to favor a democratic system of governance, free markets, and the private ownership of profits. Consumers generally seem to rely on the ability of the free-market forces of supply and demand to afford them the lowest possible prices. But as democratic governance systems don't always produce the best leaders, the market forces—though presumably impartial—are not free from manipulation and are far from being the most effective mechanisms for allocating our scarce resources.

I do not want to paint a pessimistic view of markets, the generation of profits, or justifiable remunerations for investments and associated risks. I share the view that "the remarkable power of trade and investment can be harnessed acknowledging and addressing the limitations through compensatory collective actions" and "if governments would do their job in setting up the right rules, major international companies would play a vital role in solving problems" (Sachs, 2005, pp. 357-358). This government action is essential to protect the vulnerable from the limitations of market forces and unscrupulous investors that together create situations that worsen conditions for those who are already disadvantaged, essentially kicking them when they are down. For this reason, it is worth considering the situations facilitated by the free market and the private ownership of profits that warrant some official intervention. But more importantly, these situations command the attention of consumers who benefit from the comparably low prices of goods and services produced under conditions that many deem morally and ethically unacceptable.

Unfair Wages, Poor Working Conditions, and Job Cuts

The free market does not support the social ideal of fair wages. The cost of goods and services in the free market is largely determined by the supply and/or labor cost. Disadvantaged people comprise a large share of this labor market, and many must work for meager wages just to survive. They also take great risks to migrate legally or illegally within or outside their homeland for employment. While the influx of migrants can suppress the wage rate, it can also facilitate economic growth and prosperity. Given their illegal status, some poor migrants often work in inhumane conditions and receive substandard earnings because their fear of deportation makes them feel powerless. Instinctively, some businesses take full advantage of such cheap labor by investing minimally in adequate working facilities and setting work hours that far exceed the hours set by labor regulations, increasing the likelihood of profits spiraling upwards.

It is now common practice for businesses to transfer operations to countries where labor is cheap to generate profits. This practice impacts the poor on the mainland in two related ways: directly through unemployment and indirectly by the loss of national income, which may result in a reduction in financial support for government-funded programs or initiatives that directly benefit people living in poverty. In addition, imported products manufactured in countries where labor is cheap are usually cheaper to purchase, a circumstance that often destroys domestic industries by suppressing local prices. Local industries, unable to reduce their cost of production, often cease operations, and many people, including the disadvantaged, are left without jobs.

While the living conditions of many people from the recipient countries that benefit from foreign investments may improve, others find themselves in a compromising situation when foreign companies pull out their resources at the end of tax breaks and other concessions. To illustrate, some hotels in St. Lucia often change ownership. This provides an opportunity to reset arrangements regarding tax breaks, concessions, staffing, wages, job descriptions, and fringe benefits. A change in ownership or an economic downturn, as with the coronavirus pandemic, makes it possible to dismiss long-serving employees and hire new ones (usually with less experience) at a lower pay rate. Dismissing long-serving employees and hiring new ones at a lower rate makes it possible to recover monies paid out in severance pay and other benefits. It also prevents staff members from acquiring too many financial and other benefits over time. However, those employees who are terminated or made redundant still have financial obligations, which forces some to compete and accept new or similar positions, often at a lower salary. But others are not so fortunate. Why?

Thanks to universal education, an estimated 11,000 students graduate from secondary school yearly in St. Lucia. However, only a small fraction goes on to pursue higher education. Hence, most of these students enter the labor force, increasing the supply side of the labor market. Many of them are employed by small businesses owned by foreign investors in the capital, Castries, mostly for substandard wages. However, the majority are employed in the hotel industry, and their services are key to maintaining attractive rates for vacationers. They also appear to facilitate staff layoffs and the termination of long-serving employees at the lowest possible cost. In addition, they facilitate the preservation of unfavorable working conditions since they can easily replace members of staff who decry these poor working conditions. What kinds of poor working conditions are likely to exist in the hotel industry?

Based on data from a team of researchers (Shahidsaless et al., 2018), it was found that "72% of employees at all-inclusive hotels in St. Lucia experience moderate to high stress, compared to 50% of employees at non-all-inclusive hotels" (p. 38). Workers in the hospitality sector are generally faced with a range of health-threatening working conditions and risks such as repetitive tasks, long hours, exposure to harmful substances, noise, psychological pressure, and physical aggression (Tourism Concern, 2013, p. 13). Shift work involving nights and evenings (which is common in the hotel industry) has been associated with a wide variety of health effects, including an increased risk of workplace injury, heart disease, and cancer (Institute for Work and Health, 2010). In addition, these long working hours and split-shift system disrupt workers' family and social life.

The outsourcing of services by companies with the aim of reducing costs and increasing profits also facilitates poor working conditions in some industries. To many, the outsourcing of services is synonymous with job cuts, pay cuts, and additional work duties. Often, outsourced agencies are involved in many different income-generating activities, and they deliberately lower their bids to win contracts. Regrettably, to maintain profits, both parties sometimes settle for a contractual arrangement with full knowledge that those delivering services will be adversely affected financially, physically, and emotionally.

To make matters worse, modern technology is revolutionizing the workplace and is likely to result in depressed wages and job losses in certain industries and certain segments of the labor force. Machines are more efficient and less emotional than humans; they can work for longer hours than humans, and they don't request sick days. In countries where the implementation of modern technology is slow, some entrepreneurs are likely to compensate for minimum investments in technology with low-cost labor and minimal investment in worker health and safety. It is also highly likely that laborers engaged in operations that are not highly automated will be overworked and underpaid as they aim to produce an output comparable to that of highly sophisticated automated operations. As the cost of installing modern technologies and automated processes reduces, it is apparent that there will be huge job cuts as the demand for cheap labor is likely to decline. Until then, the free market will continue to play its part in denying many the freedom to obtain decent wages under safe working conditions as they strive to live decent lives.

Sex Slavery, Child Labor, and Other Black-Market Operations

Stories and life experiences were shared at family and community gatherings in St. Lucia. This way of sharing information was a part of my life growing up, a tradition that may still be held by many, especially those in rural areas where the transfer and reach of modern technology and social media are still lagging. Many were the stories of those whose parents—out of good faith, wanting them to succeed in life—arranged for them to reside with more advantaged relatives. In return for their company and support in household chores, the host family promised to make provisions for these individuals to attend a school or gain acceptable employment to generate some income. Their parents, in return, helped in whatever way possible, mostly by supplying the host family with agricultural produce or other items they could consume or sell.

Unfortunately, some of these stories had sad endings. Many of these individuals never attended school but instead worked as slaves, performing extensive household chores, and sometimes even being forced into sexual relations with acquaintances of the host family. These experiences underscore interrelated practices that are very prevalent today, namely child/forced labor, sex slavery, human trafficking, and child pornography. These practices are fueled in part by the demand for inexpensive goods and services and the desire by some in society to satisfy their sexual fantasies, no matter the human cost. The prevalence of these activities highlights the inability of regulatory and monitoring bodies to limit the reach of market forces, creating the environment for the black market—the illegal sale of goods and services—to flourish. This situation is compounded when family members, friends, and the state are perpetrators.

The following excerpt from a 2006 report by the United States Department of State on human trafficking is sobering:

"Of the estimated 600,000 to 800,000 men, women, and children trafficked across international borders each year, approximately 80 percent are women and girls, and up to 50 percent are minors. The data also demonstrated that the majority of transnational victims were trafficked into commercial sexual exploitation. With a focus on transnational trafficking in persons, however, these numbers do not include millions of victims around the world who are trafficked within their own national borders" (p. 6).

A later report (2013) noted that "social scientists estimate that as many as 27 million men, women, and children are trafficking victims at any given time" (p. 7). The United Nations Office on Drugs and Crime reported a decline in the number of detected cases of human trafficking in 2022; however, it was also reported that though the COVID-19 pandemic limited opportunities to commit the crime, it pushed trafficking further underground and made it more difficult for law enforcement to target this activity (p. ii).

The International Labour Organization (2021) estimated that almost one in every ten children worldwide was subjected to child labor at the start of 2020, which amounted to 160 million children (p. 8). The organization described child labor as "work that children are too young to perform and/or work that, by its nature or circumstances, is likely to harm children's health, safety or morals" (p. 19). This included hazardous work, unpaid household services, and more cruel forms such as forced recruitment of children for armed conflict and debt bondage.

In 2014, officials in the United States dismantled a huge international ring sharing indecent images of children (BBC News, 2014). One of those involved in the bust said, "Never before in the history of this agency have, we identified and located this many minor victims in the course of a single child exploitation investigation." The perpetrator was operating a website, and it was reported that about "250 children-mostly boys from the US -appeared on the site." The website had more than 27,000 subscribers. Another incident that angered the public was federal authorities' arrest of a twenty-seven-year-old teacher in Massachusetts that same year, who admitted to possessing and trading hundreds of images of child pornography (Tuthill, 2014). He also allegedly confessed to "setting up and hiding a video camera in a locker room to tape children changing in and out of clothes."

Given the above, it seems like the future is grim for potential victims, as these vices are pervasive in nations that possess the capabilities and resources to monitor the behavior and act against abusers. The challenge of dealing with the exploitation of disadvantaged people for sexual gratification stems partly from society's overestimation of the ability of its members to act rationally within the limits of laws to satisfy their sexual fantasies, such as with adult entertainment. But human imaginations and desires, no matter how unconventional, cannot be easily restrained, especially once they have been triggered and satisfied to some extent. Unfortunately, some will always seek to go beyond these societal boundaries, and once there is a demand, the market mechanism will inevitably provide some equilibrium. In the short run, a measure of restraint may be obtained by heavy and extensive policing and the public's cooperation to report suspicious activities of this kind. However, the long-run solution may be the greatest work in human genetic engineering the world has ever seen or some other intervention that would diminish or extinguish the desire to exploit disadvantaged and vulnerable people for sexual gratification.

Unfortunately, there are more disturbing situations to contend with, such as the illegal harvest and sale of human organs. It is estimated that "10,000 black market operations involving purchased human organs now take place annually" and "patients, many of whom will go to China, India or Pakistan for surgery, can pay up to $200,000 (nearly £128,000) for a kidney to gangs who harvest organs from vulnerable, desperate people, sometimes for as little as $5,000" (Campbell & Davison, 2012). The following headlines from articles in the news media portray a very bleak situation: "Little girl sold for organ harvesting" (The New York Post, 2013); "They are butchering a human for money: Afghanistan's poor sell their kidneys on the black market" (Farmer & Makoii, 2021); "Black Market for Body Parts Spreads in Europe" (Bilefsky, 2012); "An organ is sold every hour, WHO warns: Brutal black market on the rise again thanks to diseases of affluence" (Gayle, 2012). Desperate for money to cover their basic needs, many disadvantaged people are making unthinkable sacrifices in a battle for their survival.

Points to Ponder

Under the preferred trade system of free markets and private ownership of profits, tradeoffs seem inevitable. These tradeoffs are often between profits and fair wages, environmental degradation and economic expansion, and cheap consumer goods and adequate occupational health and safety. Unfavorable conditions are likely to increase as companies become increasingly owned by detached shareholders who have no interactions with employees or connections to the products or services they invest in.

It is not uncommon for the public to shift the blame to multinational corporations for social and environmental problems, such as the explosion of plastic waste. However, it would be naïve to think that businesses would invest billions into political campaigns and expect nothing in return. Hence, public outrage is certainly unwarranted when some policies are tilted to favor big businesses. Sometimes we create the enabling environment for social and environmental problems by means of our purchasing power. It's fair to assume that, on average, our purchasing decisions are based primarily on price, and so, logically, the lower the price, the better the bargain. At times, the price takes precedence over our own health and safety, much less the health and safety of those involved in the commodity chain. Hence, it is logical to see that an awareness of child labor, inadequate salaries, and poor worker health and safety is limited in turning the trends in consumer spending and habits. Regrettably, people living in poverty sometimes provide the fuel to perpetuate these practices. For instance, poor people can only afford cheap clothing and food items, creating a demand for low-cost products that are sometimes manufactured under inhumane conditions. Businesses, in effect, create a system where the poor are against the poor, trapped in a vicious cycle.

The argument is often made that if consumers refuse to purchase products manufactured under poor working conditions, they are denying people living in poverty the opportunity to gain some income and the means to improve their living conditions. However, success stories may pale in significance when compared to the incidence and scope of social, medical, and environmental problems related to these poor working conditions.

Therefore, the assumption that the free-market system and the private ownership of profits are the best mechanisms to manage the world's scarce resources is wanting. For example, economists and those with purchasing power would agree that an increase in the price of a commodity induced by a shortage is the best mechanism to ensure that consumers who value the commodity most and are willing to pay the market price should receive the commodity. This means that those below a certain income level are denied access to this commodity, regardless of how much they may value it. This is the case because value judgments in the realm of the market are more a matter of the availability of disposable income than actual needs. This is partly at play in panic-buying before and during natural and other disasters. Surpluses and shortages arise in part due to the inherent inefficiency of this market model and the calculated actions of the major players, especially those with monopoly control. For example, to maintain stable oil prices and profits, some oil-producing countries have aligned forces to manipulate crude oil production, which affects fuel prices and transportation costs.

But while it is easy to conclude that the system is unfair, the probable solutions are not popular, as many believe they will stifle innovation, discourage investments, create inefficiencies, and decrease productivity. Some proposed solutions include a cap on the profits that could be generated per unit of selected product or service; the transfer of some profits directly to employees and investments in better working conditions and other areas of social/human development; the establishment of some mechanisms to protect both consumers and manufacturers against the destructive impact of a legitimate shortage or surplus; and some protection to avoid the producers of critical goods and services from closing down, laying off employees, or shifting production elsewhere due to stiff competition (providing that these operations are executed effectively and efficiently).

Fortunately, a few positive trends, such as the fair-trade movement, highlight the potential for change. The fair-trade movement empowers consumers and producers to look beyond the market price and at least consider the social, political, economic, and environmental implications of their actions. To this end, many are beginning to use their purchasing power to effect change. However, initiatives like this must be protected from those with ulterior motives who aim to derail the movement for their own ends. But are we collectively capable of devising and sustaining trade and market mechanisms that are fair, moral, and ethical? Some argue that the needed changes are biological and that because we are inherently creatures driven by self-interest, despite our best efforts, it would not take long before we would revert to this biological instinct.

Chapter 5:
Missed Opportunities: The St. Lucia Banana Industry

Today, I want to share some thoughts with you on an issue that has bothered me for some time. It is this issue-the tendency in our society to reject personal responsibility for our actions or misfortunes and to find a reason to blame someone else. We always blame others, but never ourselves.
Prime Minister of St. Lucia, 2004

The banana industry of the Windward Islands (St. Lucia, Grenada, Dominica, and St. Vincent) originated in the early 1950s, following the collapse of the sugar industry. In 1953, Geest Industries Ltd. (a UK company) and the Banana Growers' Associations signed a contract for Geest to purchase all bananas of exportable quality. The grower association in St. Lucia, the Saint Lucia Banana Growers Association (SLBGA), a statutory institution, had control of all bananas produced on the island and became the sole exporting agency for the crop.

In the mid-1990s, when Geest decided to exit the Windward Islands banana business, the Windward Islands Banana Development and Exporting Company Limited (WIBDECO) was formed and was party to a joint venture with Fyffes, an Irish company, to take over the banana business in the islands. WIBDECO was owned by the government of the four Windward Islands and their respective banana growers' associations. Consequently, the Saint Lucia Banana Growers Association Act was amended to allow WIBDECO to become the sole exporter of bananas from St. Lucia. However, the SLBGA maintained its right of ownership of all bananas of suitable quality delivered by its member growers for export up until 1998, a pivotal year in the history of the banana industry. While many claim that cheaper bananas from Latin America and the loss of preferential treatment in the marketplace contributed to the downturn in the banana industry and related problems, many internal forces worsened the situation to cause sustained poverty.

Green Gold

To date, there is no doubt that the banana industry has been and continues to be a crucial means of socioeconomic development for many St. Lucians, particularly residents in rural areas and those with limited skills. Based on available data from the Central Statistical Office of St. Lucia (www.stats.gov.lc), banana sales revenue from the UK (the main importer) totaled EC$9.44 million (about US$3.4 million) in 1966, reached a peak of EC$186.9 million (about US$69 million) in 1990, and fell to a mere EC$14.5 million (about US$5 million) in 2017. In addition, the number of farmers declined from an estimated 9,500 in 1993—a turning point in the history of the industry—to just about 160 at the end of 2019, based on banana export figures. It is believed that even fewer banana farmers were active in 2020.

The rise in income in the 1980s was an opportune time for farmers, and those who depended on them, to rise out of poverty and create systems to keep them afloat. Given the high returns from banana sales that were supporting and sustaining both big and small farmers and providing regular income, bananas were referred to as "green gold." Many farmers and farmworkers were able to build suitable accommodations, educate their children, invest in other sectors of the economy, and prepare for cases of emergency and retirement. However, some of the farmers who generated huge incomes failed to utilize these funds responsibly, due partly to a lack of business and managerial skills. Having wasted this golden opportunity, many ended up in substandard housing, heavily dependent on government support, and would never rebound financially to their previous status. In fact, a poverty assessment for 2005–2006 noted that "the rapid decline of banana agriculture has left many ex-banana farmers without the personal resources to deal with old age, in spite of the fact that in the heyday of the industry they might have acquired some of the material comforts of middle-income status" (Kairi Consultants Limited, 2006, p. 144).

The Privatization of the Industry

The agitation leading up to the 1997 general elections in St. Lucia will be forever etched in the minds of many citizens. Frustrated farmers wanting "a greater say in the running of the banana industry" (Smikle, 1996) began to take "no cut/harvest" strike actions, orchestrated by a pressure group known as the Banana Salvation Committee (BSC). It was believed that the government was responsible for the deaths of two banana farmers who were killed in 1993 as striking farmers clashed with police. Farmers became divided into two opposing groups: those who wanted to harvest their bananas and those who supported the "no cut/harvest" strike actions. In the ensuing fights, banana fields and pack houses were destroyed, and death threats were made to farmers who opposed the strike actions. The strike actions came at a financial cost to the industry, and they occurred at a time when the market environment was changing to the disadvantage of banana farmers in St. Lucia and the other three Windward Islands.

Towards the end of the election cycle, the presiding government succumbed to the demand of the BSC and granted the farmers greater control over the affairs of the SLBGA by means of majority representation on the board of directors. However, the new government that emerged from the 1997 election carried on with plans to privatize/commercialize the banana industry. Based on the Saint Lucia Country Strategy Paper (1999, p. 12), the privatization of the industry would allow for:

- an effective system of corporate administration in the industry,
- banana companies to operate in as cost-efficient a manner as possible,
- a greater sense of urgency and greater proactivity in these companies,

- a more accountable system of ownership and control, with sanctions for non-performance and rewards for performance, and
- the industry to run along strict commercial lines. In addition, it seemingly removed what has traditionally been identified as an unnecessary obstacle to efficiency—government and political interference.

The government's role in the new arrangement was "to ensure that there were clearly articulated and defined parameters within which all companies operate" and to regulate essential common services, such as pest/disease control and credit control. The government also established a new entity, the Banana Industry Trust (BIT), to "allow for the maximization of the financial resources placed in the industry" and provide a forum for policy formulation and resource management.

In September 1998, the SLBGA Act was repealed, and this statutory organization was dissolved. The assets, liabilities, privileges, and obligations of the SLBGA were vested in a new company, the Saint Lucia Banana Corporation (SLBC), which was privately owned by registered banana growers and was now a shareholder of WIBDECO. Shortly after the dissolution of the SLBGA, at least three additional banana trading companies came into existence.

These radical changes in the industry, among other things, resulted in lower returns to producers, a decrease in banana cultivation, and a consequent loss of employment. This was inevitable, as many farmers did not adjust to this new market environment. For example, the companies that emerged with privatization were comprised of people who were formerly employed with the SLBGA, some of whom had developed very close relationships with farmers. Regrettably, some farmers were supporting the ineffective and inefficient operations of the companies that emerged simply because they had grown too close to those manning these operations.

The secretary of the BSC served on the SLBC board of directors for at least two terms. In 2010, he was reelected to the board, even though he was accused of serious wrongdoing that required the involvement of the local courts. What might have led to his reelection? His remarks provide a clue: "This was a promise I made to farmers when I returned to head the association, that I would ensure that the shareholders of the company get what they are entitled to" (The Gleaner, 2011). Consistent with his remarks, the SLBC adopted a resolution for the voluntary liquidation of its assets, and not too long after, eligible farmers received an initial payment from the sale of the SLBC's shares in WIBDECO to the government of St. Lucia.

This begs the question, why did the liquidation of the SLBC occur so quickly with little or no dissenting views? Well, farmers were financially strapped, many of the former employees of the company welcomed the move since their compensation was long overdue, some financial institutions were elated since they were now able to recover the monies loaned to the SLBC, and the public seemed no longer interested in the affairs of the banana industry. With the demise of the SLBC, many of its employees found shelter in the National Fair-Trade Organization (NFTO), a producer organization that emerged from the fair-trade movement. In due time, the NFTO was the only producer organization that remained after the privatization of the banana industry.

The Fair-Trade Movement

A major shift in the structure of the banana industry took place with the advent of the fair-trade movement. Fair trade is a movement that, in principle, assures that producers in poor countries get a fair deal. This means workers' rights, safer working conditions, and fairer pay (Fairtrade Foundation). To this end, a fair-trade social premium—an amount in addition to the revenue from product sales—is paid to producer organizations. Given the poor state of the Windward Islands' banana business, shifting to the fair-trade model was presented as necessary to at least curb the large number of farmers exiting the industry. In 2009, the Fairtrade Foundation reported that "since 1992 the Windward Islands lost more than 20,000 of the 25,000 banana farmers" (p. 9).

The NFTO, a non-profit organization, started as a small producer organization under the jurisdiction of a regional body, the Windward Island Farmers Association (WINFA). The NFTO was primarily concerned with organizing farmers into groups based on geographic location and managing the distribution and use of the fair-trade social premium. However, with time, the NFTO took on more responsibilities, including the operation of the banana-buying depots on the island. These new responsibilities required competent personnel, adequate resources, and effective systems. Also, taking on these new responsibilities pushed the organization into the spotlight and opened it to more public scrutiny. Regrettably, some held the view that the structure and ethics of the organization were not sufficiently transformed to handle its operations due to the struggle by the leading members to maintain dominance, unqualified staff, nepotism, and a poor system of checks and balances. In addition, the purchasing component of the organization had taken on a life of its own, totally divorced from the primary aim of the organization—to improve the livelihood of its members. Moreover, the social premium made it possible for the organization to invest in questionable ventures and for the poor performance of the organization to go unchecked. Hence, it seemed that these combined factors countered the real benefits of the fair-trade movement.

The accusations lodged against the NFTO became increasingly difficult to refute, especially with the arrangement for the treatment of Yellow Sigatoka, a disease common to bananas. For years, many farmers reported that they were being billed for treatment applications that were never received. Moreover, they were in a catch-22: their farms were heavily infested with the disease, making the fruit unmarketable at times, but payments for these treatments were often automatically deducted from their accounts. The NFTO eventually discovered that it was being billed for services that were never delivered by the service provider. The matter reached the local courts, but although it was reported that the NFTO had strong tangible evidence (receipts, invoices, bank statements, and farmer testimonies) of the alleged act, no one was reportedly fined. It is reasonable to conclude that fully pursuing this case would have opened other members to liability, and it is unlikely that those at the top would have taken that risk. The result of this case may have emboldened others to engage in similar behaviors, as evidenced by subsequent reports of misappropriation of funds and company resources.

In 2020, there was a major change in the banana industry that required decisive actions by the NFTO. The shipping/marketing company went into administration on July 23, 2020, after defaulting on substantial payments to farmers and other creditors. Ironically, some farmers were calling on the government to have a greater say in the operations of the NFTO, essentially a cry to reinstate "what was traditionally identified as an unnecessary obstacle to efficiency—government and political interference" (Saint Lucia Country Strategy, 1999, p. 12).

In 2022, the government's efforts to resuscitate the banana industry were captured in the following statement: "We were able to assist the NFTO with getting the $3.8 million dollars to assist them with the operations. We had a banana task force established to look into the whole industry; where it is, where we want to go, and what were some of the challenges. We had a special consultancy done to review the operations of NFTO and then we moved to further reengage the farmers to ensure that they produce the quality of fruits exported to the UK market" (Auguste, 2022).

But could the NFTO be restored to its former state, and is it worthy of further investments? It was subsequently highlighted that the NFTO lacked the resources and expertise to undertake the critical role needed to secure the UK export business in the absence of

Winfresh, the shipping/marketing company (Government of St. Lucia, 2022).

Business Diversification

WIBDECO was later rebranded as Winfresh, as the company sought to diversify its business portfolio. To reiterate, this banana shipping/marketing company was owned by the governments of St. Lucia, St. Vincent, Dominica, and Grenada (the Windward Islands); the St. Lucia Banana Corporation (SLBC) was a shareholder but sold its shares to the government of St. Lucia in 2010.

At its inception, many saw Winfresh as a key player leading the banana industry in the new market environment, in which former British colonies, such as the Windward Islands, no longer received preferential treatment. Regrettably, with the passage of time, farmers and other stakeholders shared the view that "Winfresh must go" (The Voice Newspaper, 2018). Their anger appeared to be directed primarily at the chief executive and his fringe benefits. However, he was employed with the company since its inception and was still at the helm when the company went into administration in 2020, which seems to suggest that the accusations against him were not entirely valid and that most of the directors/shareholders of the company did not hold this view. Still, many farmers and other stakeholders were ready to "throw the baby out with the bathwater": to get rid of the chief executive at all costs.

Many were also concerned that Winfresh was completely swallowed by the corporate culture and that profit maximization took precedence over the interest and livelihood of local producers. It appeared that Winfresh was party to a move to push "small and unproductive" farms out of the industry in favor of "large" farms. From Winfresh's perspective, this was the only way to save the industry, in view of the radical changes in the marketplace. However, while there were potential financial benefits to getting rid of unproductive farmers, it was not a socially and politically viable option.

In addition, some of the factors that contributed to poor productivity were beyond the farmers, such as the availability of essential inputs, especially fertilizers and fungicides, on a timely basis. Moreover, with diminishing returns on investments, it was usually the less productive farmers that continued cultivating bananas, as there were no other available or viable options for them.

Winfresh also facilitated a quota/allocation system for the sale of bananas, which seemingly had no correlation with actual production. In fact, farmers were given allocations that were way below their production levels. Regrettably, there seemed to have been no arrangement between Winfresh and producer organizations to purchase and sell the additional fruit that would become available because of the occasional assistance given by the government to help cope with natural disasters.

Nevertheless, a minority would argue that, at least before 2020, Winfresh was able to weather the market forces and create some stability in the local banana industry by maintaining its position on key matters.

Many farmers still speak favorably of WIBDECO, rebranded as Winfresh, making the strategic move to trade directly with producers when it was under attack by some influential producer organizations. It was during that period that farmers who traded directly with the company claimed to have made the most substantial returns.

One of the persistent problems affecting the local banana industry is the high cost and low volume of production during certain times of the year. This problem was cited as a contributing factor to Winfresh's decision to purchase bananas from other countries where the cost of production and supply was more stable. This also made it possible to sustain the banana business in the Windward Islands, particularly St. Lucia, as the remaining three islands had either gone out of banana production or produced minimal amounts. However, many farmers harbored the view that these alternative banana markets contributed to the demise of the local banana industry.

Taking the above into account, the diversification of its business was another way for Winfresh to cushion the poor performance of the Windward Islands bananas. This brought into existence subsidiaries, including Vincyfresh Limited, Sunfresh Limited, and Grenfresh Limited. On its website (www.winfresh.net) at the time of writing, Winfresh noted that the company "invested in modern production facilities to deliver some of the freshest, tastiest Caribbean foods and drinks—bringing further income and growth opportunities to our farmers and communities."

The company saw this move as a vehicle to "growing shareholder value and continuing to serve as a wealth generator for the agriculture sector, using its resources to modernize and transform agricultural production in the Windward Islands." However, the chief executive lamented that Winfresh "needs its shareholders, suppliers, customers and stakeholders generally to believe in and support its vision and mission." Regrettably, the company never had the full support of local farmers, and to some extent the wider public, given the poor image of the company propagated by some farmers and other stakeholders. Furthermore, it was clear that these new initiatives never took off in a way to alter these deeply entrenched opinions.

In 2020, the inability of Winfresh to pay its creditors and local banana farmers did not attract the attention of the local media as one would have expected. In fact, it appeared that the only media personality who gave this matter due attention had been lamenting the plight of local farmers consistently over the years. He would often remind his audience of his family's personal hardship in the banana industry, and he attributed some of his success to the banana industry. But while it caught some by surprise, the poor financial status of Winfresh was no secret to the management team and shareholders. It was mind-boggling that there appeared to be no arrangement in place to ship and market bananas from St. Lucia in the absence of Winfresh.

And while farmers were critical of some of Winfresh's employees, in an ironic turn of events, it would be a company founded by some of these employees as far back as 2018 that would provide these services in the short run. Given the turn of events, some stakeholders still saw this company as an adversary and failed to join forces to keep the banana business afloat. Consequently, banana exports to the UK were suspended in 2022, and high shipping and fuel costs, coupled with the negative impact of past decisions by previous administrations, were cited as the contributing factors.

Financial and Technical Assistance

Banana farmers in St. Lucia have proven to be keen at exploiting gaps and inconsistencies in the systems established by the companies operating in the industry. There have been numerous reports of strange items and products other than bananas reaching the importing country in banana boxes and of farmers receiving more than one payment for the same banana consignment. Also, bananas from disqualified and/or uncertified farmers still manage to go through the system.

There are cases where farmers sold inputs that were made available through grants or on credit terms simply because there was no comprehensive system to track repayments of loans or whether these inputs were put to the stipulated use. In some cases, the farmers' acreage under production automatically increased once there was an opportunity to receive resources, but whenever the transaction involved a payment or deliverable on the part of the farmer tied to his/her acreage, the size of the farm suddenly reduced. This appeared to be the case with the assistance provided by the government to farmers after the passage of hurricanes Dean and Tomas in 2007 and 2010, respectively. This was also the case when assistance was given to deal with the Black Sigatoka disease. It stands to reason that these farmers' self-serving actions may have stemmed from their negative experiences, including the fragmentation of the industry and the struggle for the remains of what many viewed as a dying enterprise. However, if farmers cooperated or joined forces to address common problems, they would have enjoyed the financial savings and other benefits that come with sharing resources and expertise. Regrettably, farmers were yet to take the first step in that direction, to view each other as allies as opposed to adversaries.

The Windward Islands was dealt a serious blow with the loss of preferential treatment. To cushion the blow, the Windward Islands and other African, Caribbean, and Pacific (ACP) states received an aid package (Banana Accompanying Measures) from the European Union (EU). In addition, the EU had previously supported ACP countries with two successive programs focused on bananas and diversification of the agricultural sector, namely the Special System of Assistance (SSA, 1994 to 1999) and the Special Framework of Assistance (SFA, 1998 to 2008).

The Banana Industry Trust (BIT), an entity established by the government of St. Lucia, had the privilege of managing these funds. In the end, however, it appeared that contractors and consultants benefited more than the farmers, given that the use of these funds and accompanied technical assistance did not halt or stabilize the downward trend in banana production.

Points to Ponder

Sadly, the world is becoming increasingly unstable and divided socially and economically. We sometimes need someone to tell us that things will get back to normal and that the future can be secured. Further, we sometimes need someone to blame when things go wrong, as they often do. Under such unfavorable circumstances, we are more prone to become victims of misguided information and open to ideas that we would never entertain under normal circumstances. The situation becomes increasingly dangerous when the news media and influential figures in the music and film industry take advantage of our vulnerable state.

To this end, actors on opposing sides are driven by the need to differentiate themselves among the competing solutions to save humanity. Their views are often polarized, and relevant information is often distorted or released in calculated doses with the intent of misguiding the recipients to attain predictable outcomes. Therefore, individuals and institutions in positions of authority in both the public and private domains will work tirelessly to shield their policies and interventions from the attacks of their opponents and competitors. Ironically, these opponents and competitors sometimes embrace the same policies and interventions, even increasing their scope. And it is very remarkable to see how quickly people can forget the not too distant past and fall prey to the same misguided information time after time, and in the process miss out on opportunities to improve their lives.

Chapter 6:
The Justice System

The criminal justice system, like any system designed by human beings, clearly has its flaws.
—Ben Whishaw

One February morning, during the height of St. Lucia's Independence Day celebrations, one of my neighbors was allegedly killed in a hit-and-run accident. He was apparently returning home from his night job when the accident happened. He was the breadwinner for his family, which at the time included two children. The older child now felt obligated to take care of the family and had to put his educational goals on hold to find reasonable employment. In our conversations he lamented the inadequacies in the justice system and his unwillingness to pursue the matter. Sadly, his mother never fully recovered from the loss of her partner, which by all indications contributed to her poor mental health.

The operatives of the justice system would argue that its existence should serve to prevent such injustices. However, it is also accepted that the system is not beyond prejudice, as we cannot expect a perfect system from imperfect beings. Further, many people employed by the justice system are not driven by a desire to effect justice but by the need to generate income. Hence, the matter concerns how many counts of injustices we are willing to tolerate from the system. From time to time, public outrage keeps the abuses of the justice system within the socially accepted threshold. But is that enough?

Checks and Balances

Besides the need for transparency, there is a consensus that several other provisions and conditions must be in place for the abuses of the justice system to remain within its socially acceptable boundaries. A reward and sanction mechanism for those who man the system is necessary, given the risks involved and the rippling effects of abuses. Free, unbiased, and responsible media is necessary to keep abuses in the spotlight. In addition, the public must insist on the rule of law; they must ensure that poor people are not treated as second-class citizens. In the absence of these, among other things, it is very easy for the abuses of the justice system to go beyond the socially acceptable threshold. The system becomes more susceptible to exploitation by those best positioned socially, politically, and economically.

Nevertheless, many of the above-mentioned provisions exist in some countries. There is no shortage of lawyers seeking and willing to offer their services, sometimes with no down payment. The media is vibrant and bombards its audience with content in various forms. There are law enforcement officers stationed at almost every street corner, and the public is sometimes adamant that the rule of law prevails. However, in other parts of the world, the situation is very different. Sometimes the public, due to a lack of knowledge, uses emotions more than facts to assess matters of the law. Nonprofit organizations and media outlets focusing on matters of law are limited in number, and some are further restricted by their political affiliations and limited sources of income and funding. In addition, those that are less biased are often attacked and discredited by the individuals and institutions they expose.

My personal interactions with the justice system are limited but profound. Therefore, in the remainder of this chapter, I will detail a few of these experiences to highlight some of the inadequacies of the system and the consequent harm to disadvantaged people. These experiences are too limited to determine norms, but they are useful in highlighting potential abuses and their consequences.

Lessons from a Traffic Stop

I will begin with an incident involving a late-night traffic stop. The police officer requested my vehicle's insurance and registration certificate. I was subsequently accused and fined for driving a vehicle with the wrong license plates. I made several unsuccessful attempts to inform the police officer that he was mistaken. But how could an insignificant citizen correct a law enforcement officer? At the time, I had some financial difficulties and could not afford to acquiesce. I contested the charge and found myself before the courts. The officer may have realized his error since he never showed up for the court hearings, and the case was finally dismissed. There are vital lessons to be learned from this incident.

Firstly, a simple error by the police officer in assessing and gathering the relevant information resulted in an unfavorable outcome and a waste of the system's resources. Fortunately, I was innocent from the start. But there are numerous incidents where victims are denied justice because the evidence was poorly gathered, tampered with, or contaminated. Secondly, I had to request time off from work on several occasions to attend court proceedings; it usually required the entire day off work. But disadvantaged people without jobs and financial security may be unable to make numerous appearances before the court, especially over an extended period. Hence, there is a great likelihood that they may fail to make an appearance to defend themselves and be penalized for an offense they probably did not commit. Thirdly, sometimes law enforcement officials interact with the less powerful members of society in a condescending manner. This is usually the case when law enforcement officials give more significance to the power vested in the positions they occupy than to the functions they must perform.

Repeat Offenders, Prime Suspects

The second incident I wish to share will show how an innocent person can easily be prosecuted, especially if they have a history with the courts. This incident took place at the premises of a former employer. A security guard returning to the premises after leaving while on duty found that a company vehicle was missing. The police were notified, and the security guard revealed that a young man in the community, who had had some encounters with the law, was the alleged perpetrator of the crime.

The security guard recited what had happened in vivid detail, and the young man was subsequently arrested. However, it only took a few hours to discover the truth: one of the managers had taken the vehicle while the security guard was off the compound. One can only imagine the outcome for the accused man if the vehicle had really been stolen. This incident underscores how innocent people can easily find themselves on the wrong side of the law. It also shows that people with a criminal history are usually the first to be suspected and made the scapegoat for criminal incidents occurring in the communities in which they reside. This has serious implications for disadvantaged people without the resources to defend against false accusations successfully. Sometimes the premise of "innocent until proven guilty" is disregarded since some of the accused will spend years behind bars before the matter is taken to trial. It is, therefore, relatively easy to see how an innocent person can spend years in prison. Being in the company of hardened criminals can have a profound, negative impact on their view of society and jeopardize their prospects in life.

The Power of Relationships

In a third encounter with the justice system, I sought compensation for the damage sustained by my vehicle after I was involved in a vehicular accident. While I was awaiting my turn to appear before the judge, I witnessed a situation that really altered my view of the justice system. After bringing a court case to the judge's attention, an officer of the court casually reported that the files—the evidence—about that matter were missing, so they could not proceed. The judge quickly moved on to the next case. I looked for a hint of disapproval from the judge but was largely disappointed. It appeared that incidents involving the loss of evidence while in the possession of law enforcement officials were commonplace.

Tampering with evidence is very serious, as it perverts justice. This fosters distrust in the justice system when such tampering occurs while the evidence is in the possession or protection of law enforcement officials. Whatever the justification for the inability to secure the evidence, it stands to reason that some transaction was involved—financial, relational, or otherwise. Regrettably, people without financial means and strategic relationships, or just law-abiding citizens, are somewhat disadvantaged. Prosecuting those involved in the tampering of evidence is one way to build trust in the justice system. However, the time, resources, and other support needed to do so are out of reach for many, especially when certain instruments of the justice system are also the perpetrators. The 2018 human rights report for St. Lucia compiled by the United States Department of State noted that "although the government took limited steps to prosecute officials and employees who committed abuses, the procedure for investigating police officers was lengthy, cumbersome, and often inconclusive" (p. 1).

In 2020, a property under my supervision was burglarized. I decided not to make an official report at the nearby police station, given my poor experience with the justice system and the fear of retaliation from the perpetrators. However, several days after the incident, I witnessed a young man selling what appeared to be some of the stolen items. After the transaction was completed, I approached the youngster to inform him that similar items were stolen from the property. Based on our conversation, he confirmed that the items he sold were most likely stolen from the property but that he was only selling them on behalf of an associate.

Fortunately, I recorded the license plate of the vehicle involved in the transaction, and, to my amazement, a representative of the youngster utilized appropriate elements of the justice system, and the vehicle was located and the items were returned in *record time.* Unfortunately, the youngster was reluctant to reveal the identity of his accomplice. Since there was no sign of forced entry, I was very concerned that someone could enter the property as they pleased, and I feared a second robbery. Therefore, I made a formal report to the police to get their assistance in resolving the matter. To my dismay, that same representative informed me that pursuing the matter was futile, as the same elements of the justice system can be used to absolve the youngster of blame. Sadly, officers from the nearby police station approached the matter with little zeal, and the property was burglarized for a second time. This is the outcome when the relationship between citizens and the justice system is unhealthy. But all is not lost.

A few years before this incident, I was assaulted by two drunken men, and they took off with some of my personal items. I immediately reported the matter to a police officer who lived nearby, and before I arrived at the police station to make a formal complaint, the two men were already apprehended. The officer who was put in charge of the matter was very thorough, and within weeks the matter appeared before the courts. The course of events in this instance revealed that the system could be effective and the relationship between citizens and the justice system can serve a good purpose.

Overreach

The justice system can exert strong pressure on family ties, sometimes incentivizing families to break apart when difficulties arise. In times of marital crisis, divorce has become the first option for many, as securing monetary compensation through the courts seems less strenuous than repairing relationships by communicating effectively, finding common ground, and exercising restraint and humility. Regrettably, this has given rise to suicides, homicides, and broken families. The ability to utilize the justice system to secure financial support from a delinquent parent to safeguard the welfare of children is an essential provision. However, there are cases where this provision is used as a tool to place an additional burden on a parent who may truly be facing a difficult situation. Some fathers have had to dodge law enforcement personnel while in the process of seeking employment because their former partner sought to use the justice system as a tool to seek revenge for a failed relationship. Nevertheless, some capable fathers are quite comfortable having the harassment from the system be the prime motivation for supporting their children, and they contribute just what the law stipulates, no matter how meager and incongruent with the cost of living.

As human societies become increasingly divided, there is an increasing need to involve the justice system to resolve conflicts. There is no incentive for people to resolve conflicts independently; they resort to settling disputes in the courtroom. Moreover, monetary compensation is widely sought; deep and sincere apologies are no longer acceptable. Therefore, people who do not have the financial means to settle matters out of court and fear prison sentences do their best to evade the system, sometimes by migrating to countries where extradition laws are hardly enforced and local agencies are ill-equipped to deal with such cases. But disadvantaged people are usually at the mercy of the courts, and the years they spend in confinement can change the trajectory of their lives—usually for the worse. New relationships formed in prison present new perspectives and opportunities.

Points to Ponder

The justice system is built on principles to guard humanity, and therefore the written laws should at no time be an end in themselves. But the underlying principles should always take precedence, providing that those in charge of judging the principles are as objective as possible. This has proven to be true when international laws are breached to usurp despotic leaders, capture criminals, and rescue citizens on foreign soil. Therefore, when the principles that guide the apparatus are kept at the fore, it becomes a more effective tool in protecting humanity.

But we must accept that the entire apparatus is devised, manned, and monitored by rational human beings and, as such, is not immune to self-interest and the prejudice and discrimination that we all seem to have inherited as individuals, cultures, communities, or nations. Therefore, it is not unreasonable that from time to time, the apparatus will respond differently to the same stimulus, a determinant of the power and wealth of the individual, culture, community, or nation. It's not surprising that the system is willing to accept the testimony of accused criminals as evidence against others, but on the other hand, it will attack the character of past offenders to cast doubts on the credibility of their testimony. Stern sentences—death or life in prison—are imposed on individuals who commit murder, yet the law facilitates and shelters corporations and their shareholders from the damages they inflict on people on a global scale. Disadvantaged people undergo a slow and painful death due to a lack of basic needs, the cause of which can be linked directly to the unethical activities of some of these corporations.

Some powerful and wealthy people who commit crimes are released almost instantly upon making the bond payments; they hire the best lawyers and receive light or manageable sentences. Disadvantaged people can hardly afford bond payments; they generally can only afford the least concerned lawyers and thus receive harsher sentences that are sometimes inconsistent with the crimes committed. It is fair to assume that since the compensation of public defenders is not necessarily tied to their performance or the outcome of court cases, some may not be motivated to present the best possible court case. Hence, individuals who cannot afford private lawyers' services are at a disadvantage. But hiring private services does not always mean that justice will be served, as some lawyers seem more intent on exploiting the loopholes in the system to the effect that credible and incriminating evidence does not always determine the victim or victor in a legal matter.

One may wonder whether the possibility exists for the justice system to rise above all these challenges and be consistently just to all. Some propose that this can only exist when everyone is seen as equal, when wealth and status do not alter our anatomy to the extent that disadvantaged people are less human, and most importantly, when the impulse to change the status quo supplants the urge to acquiesce. Some say that it would require a global government manned by perfect beings and civilians cleansed of the inherited genes of self-interest and prejudice. Ironically, the son of the deceased man mentioned at the beginning of this chapter joined the justice system as a police marine officer. It takes a certain kind of courage to work and function effectively in a justice system that was apathetic to one's own suffering.

Chapter 7:
Education

The fundamental cure for poverty is not money, but knowledge.
Sir Arthur Lewis

Access to education that leads to decent employment has been a focal point of the discussion on tackling poverty. While universal access to education is largely desirable, the United Nations Educational, Scientific and Cultural Organization (UNESCO) reported that about 258 million children and youth were out of school for the school year ending in 2018 (www.uis.unesco.org). Consequently, the organization highlighted that an estimated 244 million children and youth between the ages of 6 and 18 worldwide were out of school in 2021. Undoubtedly, there were disparities between boys and girls and between children in rural and urban areas.

While access to education is important, it does not guarantee good performance. Moreover, many students drop out for various reasons. In 2017, UNESCO reported that 617 million children and adolescents did not reach the proficiency benchmark for reading and mathematics. It is reasonable to conclude that the situation has worsened with the advent of the coronavirus pandemic. Poor school attendance and performance are undesirable situations sustained by many factors, and I will refer to those I am familiar with and have witnessed, especially in the Roseau Valley, St. Lucia.

The Roseau Valley, located on the western side of St. Lucia, is picturesque from a distance. It captures the attention of tourists, who often stop at the designated viewing points along the main road to see a panoramic view of the area. Several small communities nestled between banana plantations (that are slowly disappearing) are captured by camera lenses and make their way to the other side of the world as visitors share photos and relate their vacation experiences on the island with friends and family abroad.

I grew up in the Roseau Valley and experienced firsthand the important role the sugarcane and banana industries had on the island's economic development. I can still vividly remember the scene on a typical Friday afternoon, the highlight of my week growing up. As soon as the siren at the old sugarcane factory sounded at 4:00 p.m., it seemed like everyone in the surrounding neighborhoods instantaneously converged in the small community in the heart of the valley. The now relatively empty pavements were covered with workers buying and selling and getting intoxicated as they spent their week's wages. The fights that eventually ensued were highly anticipated amusements.

Regrettably, this small community at the heart of the Roseau Valley, much like the surrounding neighborhoods, struggles with several social problems that impede the academic prospects of some school-aged children. The stereotype persists that people from the Roseau Valley are uneducated, ill-mannered, and backward thinking, despite the educational improvements in recent years.

Low Income and Inadequate Support Systems

A 2005–2006 poverty assessment revealed that the Anse-La-Raye/Canaries constituency, where the Roseau Valley is located (based on electoral demarcation), recorded the highest percentage of people living in poverty (Kairi Consultants Limited, 2006, p. 24). The poverty line was estimated at EC$423.83 (US$158.74) monthly, which equates to EC$5,086 (about US$1,904.87) per year (Kairi Consultants Limited, 2006, p. 19). A subsequent study in 2016 (Kairi Consultants Limited, 2018) used an annual poverty line of EC$6,433 (p. xvii), about US$2,409, and Anse-La-Raye/Canaries recorded poverty levels (headcount rate) of 38 percent, a 6.9 drop from 44.9 percent in 2006 (p. 50).

Despite living in poverty, some families invest heavily in their children's education. Some people still relate with pride that their educational achievements are the result of the hard work of their parents in the agricultural sector, especially in banana production. They express a sense of contentment as they relate their experience with assisting their parents on the farms, a situation often construed as child labor by external agencies. In such environments, however, the educational advancement of one child sometimes comes at the expense of another child. It is usually the eldest child that must sacrifice a couple of days of schooling to assist on the farm. Also, since many parents must depart very early for work or return home from work later in the morning, some children must sacrifice a few hours of school as they care for their younger siblings. Their tardiness is further compounded by the ineffective and inefficient transportation system in the area. One could imagine how difficult it would be for such children to focus on their schoolwork after expending much energy on these additional tasks. In more serious cases, some children must drop out to seek employment due to financial hardship, the poor health of their parents, and other unfavorable circumstances.

St. Lucia has a high number of single-parent households. In 2006, it was reported that "as many as 62% of children lived without one or both of their parents, with many living in female-headed households that were receiving inadequate financial support from an absent, non-custodial father" (Burke, 2006, p. 4). In 2016, a child poverty rate of 34 percent was recorded for single-parent households, and the child poverty rate was higher than the general population rate and increased considerably with the number of children in the household (Kairi Consultants Limited, 2018, p. xvii). Single parents, particularly women, face a common obstacle that makes it difficult to gain and keep employment. When family and kinship networks deteriorate, they often have difficulty finding people willing to supervise their children as they go job hunting. For those able to find employment, their salaries may be insufficient to cover daycare services. They have limited options; they must either forgo the needed employment or, if they have more than one child, require the eldest to miss school to babysit until the situation improves.

Another factor that affects educational outcomes is that children are sometimes left in the care of incapacitated relatives, particularly grandparents. In such settings, the level of care and supervision is severely compromised. This is not to say that grandparents and other relatives lack affection for these children; they do not always have the strength and resources to care for them. In an era where respect for the elderly is on the wane, many grandchildren lack sufficient enough reverential fear for grandparents to elicit obedience. These children sometimes decide when to attend school, when to come home, what company they keep, and whether they do homework. Hence, while some migrant parents send money back home, the needed guidance and support are still lacking.

Parental Neglect

I can still recall my father's inappropriate behavior when intoxicated. At times, my siblings and I had to carry him home as he was too drunk to maintain his balance. Our reward was that our peers would conveniently remind us of his behavior to keep us in check. At night, his disruptive behavior interfered with my siblings' studies and sleep.

The negligence of some parents makes it difficult for some children to attend school. Some parents have become addicted to alcohol, illicit drugs, and unsavory lifestyles. Their addictions, whether to relieve problems or for pleasure, consume much of their time and income. In addition to neglecting their education, these children are further embarrassed by their parents' unpleasant behavior when under the influence of these substances; consequently, these children are sometimes the subject of ridicule by their peers.

There are also parents who intentionally and maliciously neglect their children because of failed romantic relationships. Parental conduct in such situations will likely negatively impact their children's self-esteem and attitude toward school. The payment of child support by a delinquent parent is a provision to cushion the situation. But getting the responsible party to comply is sometimes a challenge. Moreover, the child maintenance fee, a maximum of EC$250 or US$93 per child per month (United States Department of State, 2018, p. 4), is grossly inadequate. In 2006 the maximum a parent could pay in child support then was $EC200, regardless of their ability to pay or the special needs of the child (Burke, p. 6). Hence, the child maintenance fee has remained relatively stable over the years despite increases in the cost of living.

Teenage Pregnancy and Substance Abuse

Teenage pregnancy and substance abuse (primarily alcohol and marijuana) among school-aged children affect educational prospects, as they could result in school dropout and behavioral problems. Some contributing factors to these undesired situations are poverty, lack of parental supervision, the urge to suppress painful memories and experiences, and inadequate sensitization around these issues. The easy access to some drugs and exposure to sexually charged music and movies, even on public transportation, does not help the situation.

A 2010 study by the Organization of American States (OAS) found that the lifetime prevalence of alcohol use in St. Lucia was 86.2 percent (p. 11) and the average age for first-time use was about eleven years (p. 17). It was also found that more than 25 percent of students experimented with marijuana at some point in their lives (p. 39), and the average age for first-time use was about thirteen years (p. 45). There was also an apparent strong positive correlation between marijuana use and behavioral problems in school (p. 46).

A report published by the United Nations Children's Fund (UNICEF) in 2021 on the well-being of young people in St. Lucia noted that 36 percent of adolescents from the ages of 13 to 17 have had sexual intercourse, and only 49 percent of sexually active adolescents used a condom at the last sexual intercourse (p. 7). It was also reported that as many as 22 percent of adolescents ages 13 to 17 ever used marijuana (p. 9), and 51 percent drank at least one drink of alcohol in the past 30 days, with 27 percent drinking to the point of getting drunk (pp. 8-9).

The St. Lucia Social Development Fund (www.sludf.net), an entity established by the Government of St. Lucia, listed drug use (mainly marijuana) and teenage pregnancy among the many social problems that plagued a small community in the Roseau Valley, which was also prevalent in the surrounding communities. In 2018, four members of this community "died from allegedly ingesting a toxic substance resembling strong rum" (Simon, 2018). Regrettably, school-aged children from the Roseau Valley are exposed to alcohol at a very early age. Alcohol is sold illegally in many households—an easy way to generate income for the family.

Some children born to teenage parents begin life at a great disadvantage, and their poor educational outcome is somewhat predictable. Sometimes, the fathers are irresponsible or—worse—disappear. Sometimes, they leave soon after the girl becomes pregnant or at some stage in the relationship when both the infant and the mother need them most. Their parents are sometimes heavily dependent on the limited income of poor households. They are exposed to secondhand smoke and are taught how to swear when they utter their first word. They usually spend an inordinate amount of time viewing television (sometimes unsupervised), reinforcing the toxic language and antisocial conduct in their immediate environment. In some scenarios, young parents view these infants as a barrier to enjoying their youthfulness. Hence, they leave them in the care of relatives or friends or with no supervision as they take a break from parenting.

On a positive note, some young mothers overcome the stigma associated with their pregnancy and return to school with very good outcomes. Other young parents work tirelessly to set their children on a path to educational and social success. It takes a dedicated parent to ensure that their child is ready for school before leaving for work, even if that results in the child being on the school compound hours before the school day begins. The hope is that there will be some "before-school programs" in the future to engage the minds of these children.

Government Mandates and Subsidies

Government assistance plays a crucial role in educational access and achievements. The government of St. Lucia, like most governments, recognizes the need to assist disadvantaged people in gaining access to education and improving outcomes. This is evident by the achievement of universal primary and secondary education. However, the "mandatory advancement throughout primary and secondary levels creates situations where students are promoted to higher levels in spite of lacking key skills and minimum standards in respect of literacy and numeracy" (Kairi Consultants Limited, 2018, p. 160). This means that many children will leave school without the skills needed to find decent employment.

Some school-aged children benefit from assistance with transportation, schoolbooks, and even meals. However, it is apparent that some of these provisions have a relatively small focus on disadvantaged children. For example, regarding the school feeding program, it was found that 76 percent of poor children had access compared to 69 percent of non-poor children (Kairi Consultants Limited, 2018, p. vxiii). This shows that the program benefited all students, not just those from disadvantaged households. If the same is true for the other programs, some children from disadvantaged families will likely be displaced by students from higher-income households.

The government also provides each eligible student with a laptop/tablet for scholastic purposes. However, when the home environment and supervision are not ideal, these devices are used for purposes that distract students from educational activities. At times, these devices are utilized by other family members, depriving students of the resources to conduct their schoolwork, especially those associated with virtual learning. When these devices go faulty, it is very unlikely that they will be replaced by families with limited income.

Some children from disadvantaged families have been able to benefit from government scholarships. Any financial assistance is welcomed by disadvantaged families, especially for tertiary education, as they are less likely to meet the requirements of lending institutions. But the requirements for some of these scholarships are beyond the reach of students whose circumstances have made it impossible to perform well at the secondary level. For those who do qualify, there is a downside. The downside of any scholarship program is the likelihood of people pursuing areas they are not entirely devoted to. Individuals often pursue whatever is available when they do not have the means to finance their preferred field of study. As time elapses, many develop a love for the field of study, but others only view their training and subsequent employment as a means to financial success. Hence, some people who benefit from these scholarship programs take their expertise abroad to make more money while the local services deteriorate. In this regard, there are growing concerns that the migration of healthcare workers is having a negative impact on the healthcare system in St. Lucia. Many of these healthcare workers received scholarships to pursue their studies.

New and Emerging Trends

Recently, some new trends have emerged in the community that impact educational achievements. There has been an increase in criminal and gang-related activities, and many young people are caught in the epicenter at the expense of their education. Some young people entertain the prospect of being the next millionaire, supermodel, or singer, a situation fueled by the growing number of reality television shows. Young people have a barrage of shows on multimedia platforms competing for their attention at a time when they should be immersed in homework and study, and when the home and school environment is not ideal, educational activities are more likely to take second place.

In the context of this narrative, there has also been a shift in the attitude of teachers and the public perception of the teaching profession. Traditional rules, principles, and structures related to teaching and teacher-student relations have undergone tremendous transformation in the name of modernization. I remember my adopted mother addressing a man as a "teacher," though he had left the profession over twenty years ago. Similarly, I have come across grown men and women whose children are old enough to have their own families still addressing their former teachers as "sir," whether in public or private. This reflects the level of respect that educators have earned. But that reverence for educators has declined, and some educators have developed a laissez-faire attitude toward teaching.

In summary, it takes a dedicated child, one who has a thirst for education, to remain in school when household chores and financial constraints make it difficult to attend regularly and be focused. This is further compounded when there is a lack of textbooks, electronic devices, internet access, and an inadequate environment at home to conduct assignments and studies. In addition, for some students from disadvantaged backgrounds who, from an early age, have been immersed in physical activities, sitting in a classroom for a full day of lecturing can be boring. Some, by virtue of their upbringing, tend to be vocationally inclined, and it is certainly an injustice to them to force them into an academic setting for five years just for the sake of meeting a universal secondary education mandate; while they need exposure to the associated curricula, they should not be confined to them. Moreover, the "unofficial hierarchy in the secondary school system" (Kairi Consultants Limited, 2018, p. 139) results in the assignment of poor performers in the same school and setting, which is not helping the cause. Consequently, the laissez-faire attitudes, preconceptions, and expectations of some students and teachers exacerbate the already unfavorable situation. The below-par performance of such schools and students is therefore somewhat predictable.

On the other hand, it takes a dedicated teacher and student to thrive under these challenging conditions. Teachers must endure the hostile environment in some schools where students flagrantly disrespect, verbally abuse, and physically attack them, rendering the teaching and learning process unduly onerous. The situation is indeed a complex one, and the children from the Roseau Valley must work very hard to overcome the stereotype of being uneducated, ill-mannered, and backward thinking.

Points to Ponder

Society places a high value on secular education and hopes that disadvantaged children exposed to educational opportunities will appreciate the need to perform despite all the odds against them. But several barriers make it difficult, and at times impossible, for children from disadvantaged households to take full advantage of educational opportunities.

People generally value secular education, but some need to be rescued from the toxic environment that impedes their learning outcomes—in the family, in the community, in learning institutions, and in their own frame of mind. They need an education that will not compromise content for profit. To meet that end, some need financial and institutional support, which must be protected from the unscrupulous but at the same time must not curtail their freedoms, especially freedom of speech and association. They need the assurance that the assistance provided is not under the constant threat of termination or reduction as the political climate changes. They also need to be able to navigate the systems to acquire these resources with the least possible friction, which tends to discourage some to the point where they do not even want to try. But those who are fortunate must always be aware of the greater danger: the tendency of some who have benefited from educational opportunities, sometimes with public assistance, to distance themselves from the communities and the people that remind them of their past.

Chapter 8: Violence Against Women

When women suffer, society bleeds.

"I just cannot stop worrying. I just cannot stop hustling. That's my life." These were the frequent remarks of a woman as she relayed her life experience to me. She had migrated from the south of the island of St. Lucia to the Roseau Valley in search of employment during the prime of the sugarcane industry. She was a young and naïve country girl. Her long and silky Indian hair caught the attention of one of the supervisors. No doubt, she was not the only girl who caught his attention. Wasting no time, he sought the assistance of his network of friends, who encouraged her to consider him as a potential mate. Convinced, she accepted his advances, and it did not take long before their first child was conceived. The following years were difficult times for her.

She became the subject of emotional and physical abuse in her relationship with the supervisor. He spent most of his earned income at the local bar and returned home intoxicated with nothing left to contribute to their children's welfare. He did not hesitate to insult her in public nor in the presence of their children. He often accused her of having sexual relationships with those who offered her work to gain some income to meet the needs of her children. The offer to purchase the land on which their house was built and opportunities for more meaningful employment were not his primary concern. She related with great sadness that some of the children had to leave the home sooner than expected due to the abuse. In addition, she remembered being the target of his other female friends, who ridiculed and intimidated her to the point where it bothered others in the community. Other community members who had had similar struggles then offered her advice on how to deal with the situation. She had difficulties recalling the date of their marriage since she was very sick and hardly conscious during the proceedings; since some community members felt that she would not recover from her illness, they encouraged him to marry her to increase her chances in the afterlife.

Her demeanor changed when she related how tirelessly she baked bread and made other edible products and even remembered the names of some of those who purchased these items as she passed from house to house in her efforts to gain income to take care of her children. She expressed gratitude for those who saw her plight and offered help, even with caring for the children. She also applauded most of her children for helping the best they could during these periods of distress. A sudden calmness enveloped her as she related how she cared for her abuser during his final years; despite all that had transpired, she still maintained a sense of humanity.

A Prevalent Problem

UN Women (www.unwomen.org), a United Nations entity, reported that one in three women worldwide experiences violence during their lifetime. Violence against women is referred to as "any act of gender-based violence that results in, or is likely to result in, physical, sexual or mental harm or suffering to women, including threats of such acts, coercion or arbitrary deprivation of liberty, whether occurring in public or in private life." In the Violence Against Women Reauthorization Act of 2019 passed by the United States Congress, mention is made of a more recent kind of violence against women: technological abuse. It is referred to as "behavior intended to harm, threaten, intimidate, control, stalk, harass, impersonate, or monitor, except as otherwise permitted by law, another person, that occurs using the Internet, internet enabled devices, social networking sites, computers, mobile devices, cellular telephones, apps, location tracking devices, instant messages, text messages, or other forms of technology" (p. 3). While many women manage to break free from the various forms of abuse, others, particularly women living in poverty, cannot liberate themselves for several reasons.

Often, victims are reluctant to press charges due to their reliance on the financial assistance of the abuser. The situation is further compounded when children are involved. Many women will seek help when the wounds are fresh, but as the pain lessens and the wounds heal, they are more likely to reconsider the situation and perpetuate the cycle of violence by staying in the relationship. Of note, UN Women (2022) reported that of the 81,000 women and girls killed in 2020, an intimate partner or a family member was the perpetrator in 47,000 of the cases.

The reluctance of women to follow through on reported cases of abuse can lead to cases of neglect by law enforcement agencies. Law enforcement officers may become hesitant to respond to reports of abuse, which could become fatal for women. Additionally, due to inadequate resources, law enforcement officials may not be able to respond to reports of abuse in a timely manner. For example, it was found that in St. Lucia, the lack of transportation made it difficult at times for the police to respond to reports of domestic violence in a timely manner (United States Department of State, 2022, p. 7).

It is also important to note that in some instances, the inadequacy, irregularity, and nonexistence of protection under the law may discourage some women from pursuing legal actions against their perpetrators. For example, in 2020, I followed up with a teenager physically injured by her ex-boyfriend on several occasions. In the latest incident, he had attempted to stab her, and she had sustained some wounds as she tried to defend herself. While she made a report to the local police station, she indicated that there had been no response from the police, even after several trips and telephone calls to the police station. In our last conversation, she informed me that the investigating officer advised her to seek an alternative course of action since the police could no longer pursue the matter because the time frame in which to do so had elapsed.

There are more systemic forms of violence against women. Certain forms of violence or abuse are culturally accepted and sometimes left unchallenged by the religious and political powers, possibly to avoid religious and political upsets. Women are murdered or disowned for forming romantic relationships outside their racially or socially constructed boundaries. Some are forced into early marriages, sometimes exposing them to health risks and limiting their ability to attend school. Women and girls are raped during conflict as part of the stratagems of the opposing factions. Furthermore, women in war-torn areas witness the savage slaughter of their husbands and children, leaving them in a derailed psychological state. And women and girls are trafficked within their homeland and across international borders for forced labor, including sexual exploitation.

How to Help

For women to experience the outcomes in life we all deserve as human beings, much is required of them and those who control the world's powerful organizations. This is especially true for women who have internalized and accepted their situation as the norm, as well as those who have been enslaved by a cultural construct that subjects them to cruel treatment. For example, it was found that "in 19 countries, one-third or more of women agree that a husband who beats his wife may be justified, at least some of the time" (Aizenman, 2015).

Governments and other organizations must not simply devise and ratify conventions but must take definite steps to remove culturally and socially instituted forms of violence. The public must not view such forms of violence as simple personal affairs, as we all pay the price, and the cost is twofold. Tax dollars are lost because abused women are less likely to work, and tax dollars must be utilized to create systems and facilities to aid the abused and the perpetrators. Instead of subjecting abuse victims to ridicule and gossip, incidents of abuse should be reported swiftly.

Programs should be in place to assist women in improving their employability. However, these programs and related resources must be accessible. This may involve the provision of temporary lodging and financial assistance where necessary and the decentralization of training and other forms of assistance. Women may need temporary lodging if they are fleeing from an abusive situation, as well as some starting capital to cover their basic needs until they can stand alone. They may be unable to venture too far off from their residence due to various constraints, such as finding temporary guardians for their children during their absence. To be effective, training facilities should make accommodations for children if they are seriously concerned about the welfare of their parents.

Protection from abusers and other sources of violence is also crucial. The threat of being attacked by a perpetrator can cripple a victim's ability to act; when the probability of violent encounters outside the home is high, women in abusive relationships do not have a strong enough impetus to flee. It would be of little comfort to women who flee from abusive relationships to be subjected to other forms of violence and discrimination outside the home, such as sexual harassment and restrictive laws and views in the workplace that limit employment, promotion, and equal compensation. The abuse of women during periods of war and civil unrest, on the other hand, will need a more ingenious intervention that will abate the innate desire for supremacy and curtail the demonic means and methods of fulfilling this urge.

Moving Forward

Nevertheless, some women have taken the first step towards living a normal life, and they need strategic support from the public and private sectors to sustain their efforts. A growing number of women in St. Lucia are venturing into areas of work traditionally dominated by men. St. Lucia has been featured worldwide as an attractive tourist destination. To gain a share of the "tourist dollar," young men would position themselves on popular routes to sell ripe bananas to tourists. Often, the quality of these bananas was inferior. However, this activity, which appeared to have been dominated by males, underwent a drastic change that typifies the initiatives and actions women can take to improve their status.

Along a popular route leading to the Pitons, a UNESCO World Heritage Site located in St. Lucia, a young lady entered this roadside business and created the thrust for other females to follow. The product she offers, bananas, is not only of superior quality but is well presented. The downside, however, is that many other female vendors initially sought to imitate her exact product and set up stalls along the same route instead of developing a different product that would add value to the route and put everyone in a better position to succeed. Some vendors developed a totally different product line, which is truly commendable.

Regrettably, the ability of women to take advantage of employment opportunities in the hotel sector was dealt a serious blow by the coronavirus pandemic. Roadside vendors generally catered to visitors who arrived primarily by sea, and many have had to close because of the halt in cruise ship travel and vacations. In addition, the closure of most of the hotels on the island meant that many workers, including women, were either laid off or terminated. Many of these vendors and workers were relieved by the increase in tourist arrivals in 2023.

There has also been a shift in the security services industry. Security guard positions, traditionally held by males, are now being filled by an increasing number of females. Women are now poised to dominate this field of work. This further highlights the steps that women are taking to improve their social status. The extent and reach of these changes will depend on the impartiality of employers and the willingness to challenge traditional roles and norms. Also, it is very important for these women to be duly compensated, as this will attract more women to this field of work.

But these forms of work also come with new challenges. A great threat to workers' health and safety, especially those working the night shift, is the possibility of rape and ambush. In developing countries like St. Lucia, transportation services outside the capital, especially at night, are inadequate or nonexistent. While many employers have been considerate in providing nighttime transportation, it sometimes covers only a fraction of the journey. Thus, it is not uncommon to see workers, some of whom must return to work early on the following day, hitchhiking their way home or even making the rest of the journey on foot. Also, due to a lack of parental supervision, especially at night, the risk of children engaging in questionable behavior is higher.

Points to Ponder

The role of women in the social and economic spheres of life is invaluable. Women not only play a key role in parenting and the overall nurturing of their families, but their economic contribution beyond the home is growing as more women enter the workforce. More women are penetrating previously hard-to-reach places. They are slowly but steadily gaining the freedom to decide which relationships to pursue and set the conditions. Also, many women are now the primary breadwinner for the family. It appears that the calls to end the conditions that fuel violence against women are having some results after all. However, these successes may be limited to women of relatively higher social status, those from specific geographic locations, and those from less restrictive cultures and religions. However, the balance has yet to be restored. For example, while the government of St. Lucia recently amended the twenty-seven-year-old Domestic Violence Act to make it more attuned to the challenges faced by women, the country has also embraced a local genre of music that generally portrays women as sex objects. Nevertheless, the fear of emerging and persistent problems should not inhibit us from continuing to improve on solutions to this significant social problem.

Chapter 9:
Older Adulthood

Gray hair is a crown of splendor…
Proverbs 18:31 (New International Version of the Bible)

By 2050, it is projected that "one in six people in the world will be over age 65; the number of persons aged 80 years or over is projected to triple, from 143 million in 2019 to 426 million in 2050; and the share of the population aged 65 years or over is projected to double between 2019 and 2050 in the Caribbean region" (United Nations, 2019). While these estimates may have to be revised considering the COVID-19 pandemic, they highlight a projected jump in the number of older adults worldwide.

As we age, we commonly experience declining biological, physical, and cognitive functioning. Our declining immune systems make us more susceptible to infections such as the coronavirus, which is killing older adults at a disproportionately high rate. While growing old is a necessary part of life that can bring on many adverse challenges, the negative aspects of aging can be minimized by complementary or compensatory services and resources. But while some older adults have access to these resources and services, others are at a great disadvantage.

In some societies, especially in developed countries, aging is a blessing to businesses in many ways, as people try to delay the process and find ways of coping with new challenges. Older adults are the tenants of senior homes/communities and other related facilities; they consume a fair share of drugs, supplements, and recommended foods; they undergo numerous surgeries (cosmetic and medical); they are the subject of research and technological advances; and they are often the targets of scams and frauds. Without a doubt, aging makes possible a reliable and socially acceptable market in which to be exploited. In the

United States alone, it is estimated that services and products for older adults now constitute a considerable share of the economy and are poised for further growth given the projected increase in the number of older adults (Sightings, 2013).

Children and caregivers of older adults may also view aging as a blessing. It presents an opportunity to either sever or fortify relationships and commitments. The availability of services for seniors presents a socially acceptable avenue in which to sever biological ties. Children and caregivers who are financially able can simply transfer their caretaking responsibilities to someone else. On the other hand, some seniors are the subject of outstanding care by children and caregivers as they endeavor to spend as much time as possible with them. Aging is also a blessing for those who are financially well-off and capable of deciding where to spend their final years, whether at a senior living facility or some other suitable accommodation of their choice.

Religious and political organizations also benefit from older adults. Usually, religious convictions grow stronger as we age, and by the time we retire, our political choices are stable. Therefore, older adults can be counted on to attend religious services and give liberally to institutions. During election times, politicians make every effort to ensure that seniors are reached and their votes are counted. Additionally, scientific and research institutions receive public and private funding as they conduct research and experiments on improving longevity/life expectancy. For some, however, aging can be a burden. This is especially true for disadvantaged people.

Burdens of Caring for Older Adults

I remember growing up with a reverential fear of the older population. I would never address an older adult by their first name, let alone ridicule them. They had the authority to discipline when necessary, and no child would dare question that authority, for to do so would result in further chastisement. They also loved the children in the community, much like their own, and would provide the necessary care for these children when their biological parents needed help. In addition, they possessed a wealth of knowledge, particularly on traditional medicine (from which I've personally benefited). However, it is not uncommon now to witness verbal, physical, and emotional abuse of older adults. Of note, poor public knowledge of age-related conditions, like dementia, contributes to such abuse.

Regarding the status of some older adults, the prime minister of St. Lucia in 2012 had this to say at the launching of a support program:

> "We now are faced with weaker family support structures for our aged persons, with many persons over the retirement age being forced to continue to work to support themselves. Of course, there is also the desire for persons to keep active in their twilight years. Many parents have lost their children who have been compelled to migrate in search of employment opportunities abroad. Many of our elderly live alone with no one to care for them. It does not help that our society too is becoming increasingly uncaring. Many children believe that the best places for their parents are the homes for the elderly, whether public or private. Some 'dump' their parents in these homes and never find the time to see them, comfort them, and share memorable moments of their shared past. It is a behavior we must repudiate."

While I agree that we must repudiate the inhumane behaviors directed toward older adults, we must also understand what contributes to these behaviors—especially the perceived or real burden of caring for older adults. As their children and relatives migrate, whether locally or internationally, in search of jobs and skills, older adults are often left alone. Some of them receive assistance from their relatives; some do not. For some of these migrants, time away from home is a long-awaited relief from the "burden" of caregiving. Ironically, some older adults grow in importance after their death. While money might not have been readily available to care for them, families and friends often provide funds—even beyond what is needed—to conduct the best funeral services possible.

Older adults can become a burden to some disadvantaged families as the cost of caring for them becomes more than they can bear, especially when older adults can no longer provide services for the family that would free up time for working family members. As people age, there comes a time when they can no longer supervise grandchildren (or other relatives) and perform household duties. This increases the workload on caregivers who are employed outside the home and reduces the amount of income that is available to spend on the family.

In addition, when older adults can no longer perform activities of daily living such as eating, bathing, and toileting, caretaking can become a burden, especially if the responsibility falls on one family member. It is usually the family member who is the most empathetic, who lives with the older adult, or who lives in the closest proximity who bears the brunt of caregiving. It is said that when people are under tremendous pressure and stress, they are likely to think and act irrationally. Hence, the pressure and stress associated with being a sole caregiver can be overbearing, and this situation can contribute to the problem of elder abuse. Because of the situation that some older adults find themselves in, they lose their decision-making power and have no control over when to eat, bathe, sleep, and perform other basic functions.

Older adults can also become a burden for religious organizations and state institutions, as the cost of providing care and emotional support can outweigh their contributions. With aging populations, governments may face challenges in allocating resources between the needs of the young and old.

In developing countries like St. Lucia, older adults who want to remain active are limited due to the inadequacy of building/architectural designs and transportation. Currently, the transportation system does not accommodate older adults, and some passengers are sometimes impatient with those who struggle to travel on public transportation.

Happily, the government of St. Lucia has established a program that provides older adults with home-based assistance. There are also efforts by local organizations in collaboration with the government to provide food and housing for older people living in deplorable conditions. But much more is still required for older persons to feel pride in their "grey hairs."

Points to Ponder

Despite the improvements in longevity and overall quality of life, aging is a reality that we must confront no matter how much we attempt to mask the visible signs, such as wrinkles, baldness/thinning of the hair, and declining physical functioning. It is heartbreaking to know the extent to which people will expose themselves to harm by undergoing cosmetic surgeries and ingesting various drugs and other products rather than simply accepting their aging bodies. It is, therefore, easy to see why many people are so preoccupied with these matters. This does not mean that we should take a lax approach to our health and appearance, but we need to be realistic. There are matters of more serious and universal concern.

Given that aging is inevitable, it is the course of wisdom to prepare (psychologically, physically, and financially) to deal with it. In fact, those who are better positioned do just that: They take advantage of investment opportunities, purchase adequate health and life insurance, seek employment with organizations with satisfactory retirement benefits, take advantage of state-funded or sponsored initiatives, and ensure that their children will be financially able to care for them should the need arise. In sum, they explore and take advantage of all opportunities to ensure a better future for themselves. They look forward to not having to report for work, spending more time with family and friends, traveling, and undertaking many more leisurely activities.

However, many are fully aware of the calamitous time that awaits them but are not in a position to build their defense and coping mechanisms. While aging and death are natural, the poor social, economic, and health status of some older adults are certainly not. This unfavorable situation is naturally worse in poor countries with limited resources. Some people just cannot put aside any savings for the future, as their income can hardly meet their immediate needs. Since many may be employed as "unskilled" laborers, they have no retirement program. They may find themselves having to work throughout their lifetime. While it is well documented that healthy eating habits early in life can result in huge health benefits in later life, disadvantaged people usually can only afford the cheapest and least healthy foods. Sadly, as many suffer later in life due to these factors, they may turn to suicide as a way out.

Chapter 10:
Religion

Nowhere has the prosperity gospel flourished more than among the poor and the working class. Told that wealth is a sign of God's grace and favor, followers strive for trappings of luxury they can little afford in an effort to prove that they are blessed spiritually.
Cathleen Falsani

A discussion of poverty would only be complete with considering the impact of religion. Religion can bring out our deepest emotions and cause us to act in a manner deemed appropriate by those who trigger and harness those emotions. Some religions promote a simple way of life, while many of their advocates act contrary to this principle. However, some more modern religionists are forthright and sermonize the accumulation of wealth as the "Christian" way of life, and their own lives highlight the pinnacle that their congregants should aspire to. I attempted to find a concise meaning for this fundamental teaching, referred to as the "prosperity gospel." One definition is simply that God blesses those he favors most with material wealth. However, one Guardian newspaper columnist expounded on this definition: "What they teach is that God loves and rewards his faithful servants with earthly goods. In practice, the way to do this is to shun the company of losers and hand over 10% of your earnings to the pastor. The riches he accumulates are then a sign that God has blessed him, which means that he is capable of passing on these mighty blessings to you" (Brown, 2013).

The investigation of several religious organizations in the United States regarding the possible abuse of their tax-exempt status is worthy of note. In 2007, a United States Senator announced an investigation into the tax-exempt status of these ministries, asking them to divulge specific financial information to determine whether ministry heads inappropriately utilized funds collected by these organizations. These instances included "the personal use of Rolls Royce cars, private jets, and multimillion-dollar homes." Other areas of concern were "exorbitant salaries, so-called 'love offerings' or cash payments to ministers; a justification for layovers in Hawaii and the Fiji Islands; and in one case, the purchase of a $23,000 commode with a marble top" (Lohr, 2007).

While there appeared to be no conviction in this instance, a church leader in London was ordered to repay "£200,000 after it emerged that he used church assets to buy a £13,000 Florida timeshare and spent £120,000 on his birthday celebrations, including £80,000 on a car" (Brown, 2013). A 2011 article in Forbes Magazine highlighted the five richest pastors in Nigeria, with the number one rank having an estimated net worth of US$150 million (Nsehe, 2011). Another article in 2013 indicated that the richest pastor in Brazil was worth an estimated US$950 million. The article said that "he is continuously involved in scandals, mostly due to allegations that his organization had siphoned off billions of dollars of donations intended for charity. There have also been official charges of fraud and money laundering" (Antunes, 2017). Sadly, in 2020, five people were killed in an attack on possibly one of the largest and richest churches in South Africa. It was reported that the attack was probably motivated by a feud between church members. One news report indicated that "five rifles, 16 shotguns, and 13 pistols, along with other weapons, were found at the church." It also indicated that "the church's leadership has reportedly been the subject of infighting since its former leader died in 2016" and that "the year before, the church's finances had come under the spotlight, amid allegations some 110m rand ($6.5m; £5.2m) had gone missing" (BBC News, 2020).

There are usually cases of abuse and misuse whenever and wherever money is involved, even under the strictest regulations and supervision. This is even more so under the apparent lax system of accountability, which has been reported in some religious organizations where members must exercise faith and trust. The abuse or misuse of the tax-exempt status of these organizations is a matter for the state to address, but I wish to highlight the possible impact of this theology, the prosperity gospel, on people living in poverty.

This theology seems more like an attempt to offset the contradictions between the desire for self-indulgence promoted by the free market and the need for self-sacrifice promoted in the religious text they purport to live by. It relieves those with the desire for this self-indulgence of guilty consciences if they also undertake acts of self-sacrifice by giving away some of the wealth they have acquired. If "God" blesses them with such material assets, it really does not matter under what conditions these assets were acquired, inhumane as they may be, since "God" works in mysterious ways. Who dares to question "God's" actions and deny his blessings? In essence, people are free to entertain their own material interests with no guilt once they meet their financial obligations to these organizations, which "God" sees and showers them with further blessings. What is the apparent situation of disadvantaged people who embrace this theology?

It is beneficial to divide disadvantaged people who embrace this theology into two categories. One category includes those who are genuinely spiritual and believe their poverty is a sign of "God's" disapproval, so they selflessly practice giving, pray fervently, and wait for his material blessings. However, material blessings will be indefinitely postponed for some, as the market mechanism will prove to be more powerful than the "God" they worship. But they are willing to wait as their hopes and emotions peak during church services, even though they remain at a low ebb for most of the week. Further, the handout they receive from the charitable arm of these organizations reminds them that they have not been entirely forgotten by "God" and that they need to be more trusting and faithful. Perhaps the postponement of their blessings may be a direct consequence of the amount they are giving, particularly in financial offerings. And it is certainly a frightening predicament to be in a position where one genuinely cannot give much financially.

The second category includes those who view these organizations as a means of satisfying their immediate material needs. Their primary concern is to benefit from the philanthropic arm of these organizations. They attend every church gathering to legitimize their claims and vehemently defend the faith. No revelation of misconduct or abuse can alter their devotion.

In small economies like St. Lucia, where the number of wealthy adherents is limited, disadvantaged people are almost guaranteed membership. Pennies accumulate to dollars, and the greater the mass of penny givers, the more dollars can be accumulated. But while disadvantaged people may benefit from the philanthropic arm of these organizations, they may be reminded of their poverty and their disapproved state before their "God" every time they attend mass; each session reminds them that "God" has failed to shower them with as many material blessings as their brethren have received. But disadvantaged members still consider themselves in an advantageous position, believing things will be better in the afterlife.

Points to Ponder

People generally look to religion to provide solace and hope, especially when they are hemmed in on all sides by life's challenges. They look toward their brethren to treat them as equals, as fellow worshippers, and not to judge them based on their poor socioeconomic status. They wish to be treated as humans and not as objects for exploitation. To this end, they are willing to look beyond the atrocities against humanity in the name of religion, which instill in others abhorrence for all forms of religion. But this hope held by disadvantaged people can only materialize if religious leaders have truly been cleansed of the innate desire for wealth and power, leading others to commit gruesome acts against humanity. Regrettably, it seems that the redemption process cannot erase the indelible stain of self-indulgence, which still overtakes many religious leaders today as it has in the past.

Some religious leaders deliver sermons intending to instill reverential fear for "God" in their audiences but fall short of instilling that fear in themselves as they exploit their brethren and abuse their office of oversight for material gains. To legitimize their actions and claims, they encourage their brethren to take full advantage of the political and economic systems, though they often refer to these as part of an ungodly system awaiting destruction. But the harsh reality is that the prevailing economic system is based on trade-offs, winners, and losers. Only a selected few will get to experience "heaven on earth."

Conclusion:
No End in Sight!

Without reflection, we go blindly on our way, creating more unintended consequences, and failing to achieve anything useful.
–Margaret J. Wheatley

How should we deal with poverty? This question confronts almost every generation. It presents the opportunity to search through the past and assess the present to chart a better future. But what have we learned from the past, and how can current situations inform us on the way forward? The following are just a few observations that will determine whether we take a leap of faith into the future or we approach it in a realistic and pragmatic manner.

Emotional Poverty

The definitions and measures of poverty focus overwhelmingly on material needs. Hence, it would be beneficial to broaden the definition of poverty to include not only a lack of the basic material/physical needs, but also a lack of emotions, such as empathy and compassion, that contributes to the large number of people worldwide who lack access to their basic needs.

While the coronavirus pandemic has irritated enduring wounds, such as racial, health, and economic disparities, it has also brought this emotional poverty to the fore. While some families benefit from the economic opportunities made possible by the pandemic, other families suffer the loss of their loved ones. While some countries have a surplus of vaccine doses, other nations are lagging way behind. It is said that the high number of unvaccinated people creates the conditions for the virus to mutate and that more mutations are expected once the vaccination rate remains the same. Regrettably, past negative experiences with health interventions are preventing many from taking the vaccine, while others are influenced by various theories that discredit the vaccines and question the motives of scientists, vaccine manufacturers, and even governments.

But when poverty is broadened to include the lack of emotions like empathy and compassion, the prevalence of poverty becomes less confined to poorer nations and people of lower socioeconomic status. Perhaps this will give momentum to efforts not only aimed at removing the barriers that prevent access to the basic needs but also to the psychological state that allows for the creation of these barriers.

Weak Consensus

As society tries to grapple with persisting and emerging problems, the experts will continue to devise ideologies to inform the way forward. It would be beneficial and cost-effective if the experts and world leaders would agree, but this is not always the case. However, there are some glaring observations that are indisputable. The prevalent and persisting poverty in even the richest nations and fastest-growing economies indicates that economic growth without distribution or "redistribution" of opportunities and wealth will not yield satisfactory results. Structural adjustments and austerity measures do not always deliver and can sometimes worsen the outlook for people living in poverty (due to reductions in social spending, among other things). We have also learned that development cannot be sustainable without commitment from all parties, from the least powerful to the most powerful, and that development cannot be a top-down process. Therefore, the knowledge and lessons we learned from past attempts to deal with poverty should inform future initiatives. However, given the multiplicity of actors and the complexity of the problem, consensus on the way forward has always been unstable, incoherent, and inadequate.

The Fallacy of the Free Market

If we subscribe to the economic theory of scarcity, then we must submit to some form of planning to ensure the equitable distribution of scarce resources. As such, we cannot rely on the free market to make the best decisions for us regarding the use and distribution of our natural resources and byproducts. We are much too complex to allow ourselves to be reduced to such simplistic organisms being guided by that which is abstract and impersonal. As emotional and intellectual beings, we should not always settle for and respond to the prices or conditions set by the market, as proponents want us to believe, but we must see the other elements attached to these prices, such as human suffering and the destruction of the environment. We should not be so gullible as to believe that our happiness depends on satisfying our personal needs with no relation to our fellow humans and the environment. Moreover, we should never allow ourselves to be deceived into believing that prices are really set by the market, since some players exert undue influence on market forces and benefit from special privileges. Additionally, there is even greater danger when governments and other regulatory/monitory bodies are too closely aligned with overly powerful market players that may cloud their decisions and allegiances. And for these reasons and more, we should be prepared to pay higher prices when the market is fair and those at the bottom of the commodity chain are duly compensated.

The Destructive Nature of Power and Authority

It takes a certain kind of humility to be in reach of a powerful position and not seize it at once. It takes a certain kind of self-control to be in a position of power and not abuse it. It also takes a certain kind of loyalty from those in positions of power to resist those who try to influence the way they wield their authority. It further takes a certain kind of courage to relinquish power when it becomes necessary to do so. Sadly, these kinds of humility, self-control, loyalty, and courage are rare. This is the reason why some leaders amend the constitution of their nation to extend their rule, while others do so by force. It is also the reason why the leaders of some social movements make further demands when their initial demands are met. Hence, we should be very cautious when placing unwavering faith and trust in those in positions of power, especially those at the helm of our social, economic, political, and religious institutions.

Because the human qualities that foster good governance are rare, what sometimes emerges are systems of governance that are detached from those who are governed. Therefore, it is obvious that the interest of some people, especially the less powerful, will not be adequately pursued. This is further compounded by an election and selection process that sometimes creates a chasm between friends, families, employees, worshippers, and neighbors. This usually results in a less cohesive society that breeds and sustains poverty.

Once Bitten, Twice Shy

In times of crisis, disadvantaged people are called upon to support the toppling of despotic leaders and regimes, pick up arms and defend their country, exercise their right to vote, join social and political movements to effect change, and do their part to save humanity. However, while the contribution of disadvantaged people may be essential to achieve social, economic, and political aspirations, things that really matter to them are often placed on the backburner when those battles are won, or they are simply abandoned when victory is elusive. Hence, it is not surprising that disadvantaged people are at times suspicious of certain institutions, programs, and movements, even those that are genuinely in their interest.

Burdening the Youth

The younger generation is always burdened with fixing the problems of past generations and securing the welfare of future generations. However, there are some trends that seem to render the prospect of young people to effect change improbable. It seems logical to suggest that those who are most affected by world problems are naturally those most committed to addressing them. However, it appears that they are ill-equipped to effect change since they are the least likely to attend school, complete school, and gain access to institutions of higher learning. Sadly, many young people develop their worldview from social media platforms, which often mislead them and impact their ability to deal with real problems in a realistic way. In addition, those who do listen to news media outlets are also at a disadvantage, since many media houses focus too much attention on improving and maintaining their ratings to the extent that they are willing to distort the truth and provide analyses that are geared more towards titillating the public than educating them.

Young people also devote a lot of time exploring the latest technologies and building their virtual personas; therefore, for those who are unable to exercise self-restraint, the time spent on these endeavors is likely to eclipse the time devoted to scholastic pursuits and the acquisition of life skills. On the other hand, educational institutions that place too much emphasis on their ratings (which are partly dependent on students' pass rates) sometimes compromise the integrity of their programs to increase students' pass rates. Hence, the training and ethics acquired by students who favor such academic settings are sometimes detached from reality.

While future generations have attempted to address some of the problems of past generations, the footprints of these problems are still around and continue to have debilitating consequences. Moreover, the current problems that the world faces appear to be greater in scope, reach, and impact, all happening within the context of a socioeconomic and geopolitical system where people are becoming more self-centered—some by choice, others by force. This begs the question: is there sufficient basis to place such confidence in the ability of future generations to solve the world's persistent problems, including poverty? And do we have the right to impose such burdens on them?

Bibliography

2018 Country Reports on Human Rights Practices: Saint Lucia. (2019, April 29). Retrieved October 30, 2020, from https://www.state.gov/reports/2018-country-reports-on-human-rights-practices/saint-lucia/

Aizenman, N. (2015, March 18). *Alarming number of women think spousal abuse is sometimes OK*. NPR. https://www.npr.org/sections/goatsandsoda/2015/03/18/392860281/alarming-number-of-women-think-spousal-abuse-is-sometimes-ok

Antoine, A. (2021, February 24). *Ministry of Agriculture hands over predial larceny safety gear.* Saint Lucia—Ministry of Agriculture. Retrieved May 5, 2023, from https://agriculture.govt.lc/news/ministry-of-agriculture-hands-over-predial-larceny-safety-gear

Antunes, A. (2013, January 17). The Richest Pastors in Brazil. Retrieved March 8, 2015, from http://www.forbes.com/sites/andersonantunes/2013/01/17/the-richest-pastors-in-brazil/

Arsenault, C. (2012, September 20). *Melting Arctic heats up resource scramble. Retrieved January 23, 2015, from http://www.aljazeera.com/indepth/features/2012/09/20129161337174516 22.html*

Auguste, N. (2022, August 30). *Efforts made to reignite the banana industry*. Saint Lucia—Office of the Prime Minister. Retrieved April 13, 2023, from http://opm.govt.lc/news/efforts-made-to-reignite-the-banana-industry.

Banana Company to enter liquidation. (2011, January 28). Retrieved April 13, 2023, from https://jamaica-gleaner.com/gleaner/20110128/business/business92.html

Ben Whishaw Quotes. (n.d.) Retrieved October 30, 2020, from https://www.brainyquote.com/quotes/ben_whishaw_517375

Bilefsky, D. (2012, June 29). *Black market for body parts spreads among the poor in Europe.* The New York Times. Retrieved January 10, 2022, from https://www.nytimes.com/2012/06/29/world/europe/black-market-for-body-parts-spreads-in-europe.html

Brown, A. (2013, May 29). The prosperity gospel makes a mockery of Christianity. Retrieved March 8, 2015, from http://www.guardian.co.uk/commentisfree/andrewbrown/2013/may/29/prosperity-gospel-mockery-christianity

Campbell, D., & Davison, N. Illegal kidney trade booms as new organ is 'sold every hour' (2012, May 27). Retrieved February 10, 2015, from http://www.theguardian.com/world/2012/may/27/kidney-trade-illegal-operations-who

Cornibert Must Go! (2018, December 01). Retrieved November 10, 2020, from https://www.thevoiceslu.com/2018/12/cornibert-must-go/

Davies, A. (2013, May 30). *Elon Musk: We need to leave Earth or risk extinction.* Business Insider. Retrieved May 7, 2023, from https://www.businessinsider.com/elon-musk-leave-earth-or-risk-extinction-2013-5

Encyclopaedia Britannica, inc. (2023, April 11). *Black death.* Retrieved April 26, 2023, from https://www.britannica.com/event/Black-Death

Esty, D. (n.d.). *A Term's Limits. Retrieved January 25, 2015, from http://www.foreignpolicy.com/articles/2001/09/01/a_terms_limits*

Falsani, C. (n.d.). The prosperity gospel. Retrieved March 8, 2015, from http://www.washingtonpost.com/wp-srv/special/opinions/outlook/worst-ideas/prosperity-gospel.html

Farmer, B., & Makoii, A. (2021, February 23). *'They are butchering a human for money': Afghanistan's poor sell their kidneys on the black market.* The Telegraph. Retrieved April 13, 2022, from https://www.telegraph.co.uk/global-health/climate-and-people/butchering-human-money-afghanistans-poor-sell-kidneys-black/

Gayle, D. (2012, May 28). *An organ is sold every hour, who warns: Brutal black market on the rise again thanks to diseases of affluence.* Daily Mail Online. Retrieved January 10, 2022, from https://www.dailymail.co.uk/health/article-2150932/An-organ-sold-hour-WHO-warns-Brutal-black-market-rise-thanks-diseases-affluence.html

Government of St. Lucia (2004, October 4). The Blame Syndrome. Retrieved February 22, 2015, from http://archive.stlucia.gov.lc/primeminister/former_prime_ministers/kenny_d_anthony/conversations/2004/conversation_october_4_2004.htm

Government of St. Lucia. (2008). Ministry of housing, urban renewal and local government. St. Lucia National Housing Policy

Government of St. Lucia. (2012, June 18). Remarks by Prime Minister Hon. Dr. Kenny D. Anthony at the Formal Launch of the National Initiative to Create Employment (NICE). Retrieved January 26, 2015, from http://www.govt.lc/www/text/address_anthony_18June2012.pdf

Government of St. Lucia. (2022, March 30). *Budget statement on the estimates for revenue and expenditure for the fiscal year 2022/23 by Hon. Philip J. Pierre prime minister and minister for finance, economic development and the Youth Economy March 29, 2022.* Saint Lucia—Access Government. Retrieved April 22, 2023, from https://www.govt.lc/news/budget-statement-on-the-estimates-for-revenue-and-expenditure-for-the-fiscal-year-2022-23-by-hon-philip-j-pierre-prime-minister-and-minister-for-finance-economic-development-and-the-youth-economy-march-29-2022

Government of St. Lucia. (2022, November 2) *Minister for Agriculture Issues Statement on UK Banana Trade Suspension.* Retrieved April 24, 2023, from https://www.govt.lc/news/minister-for-agriculture-issues-statement-on-uk-banana-trade-suspension

Haynes, R., Astaphan, J., Johnson, B., & Springer, C. (2004, September 5). *Report of the Health Review Commission.* Retrieved April 22, 2023, from https://archive.stlucia.gov.lc/docs/HealthCommissionReport.pdf

Hedges, S., Cohen, W., Timyan, J., & Yang, Z. (2018, November 13). Haiti's biodiversity threatened by nearly complete loss of primary forest. Retrieved October 30, 2020, from https://www.pnas.org/content/115/46/11850.

Holden, K., Charles, L., King, S., McGregor, B., Satcher, D., & Belton, A. (2015, December 22). *Reducing health disparities and improving health equity in Saint Lucia.* MDPI. Retrieved May 4, 2023, from https://doi.org/10.3390/ijerph13010037.

International Labour Office and United Nations Children's Fund. *Child Labour, Global Estimates 2020, Trends and the Road Forward.* (2021). ILO and UNICEF. Retrieved May 13, 2023, from https://www.ilo.org/wcmsp5/groups/public/---ed_norm/---ipec/documents/publication/wcms_797515.pdf.

Kairi Consultants Limited (2006). *St. Lucia Country Poverty Assessment Volume 1: Main Report.* Retrieved February 22, 2015, from http://archive.stlucia.gov.lc/docs/AssessmentOfPovertyInStLucia/Assessment_of_Poverty_in_St_Lucia_Volume_1_Main_Report.pdf

Kairi Consultants Limited (2018, December 7). Saint Lucia National Report of Living Conditions 2016. Retrieved November 14, 2020, from https://www.stats.gov.lc/wp-content/uploads/2019/01/Saint-Lucia-National-Report-of-Living-Conditions-2016-Final_December-2018.pdf

Lavigna, R. (2014, November 28). *Why government workers are harder to motivate*. Harvard Business Review. Retrieved April 13, 2023, from https://hbr.org/2014/11/why-government-workers-are-harder-to-motivate

Lohr, K. (2007, December 4) Senator Probes Megachurches' Finances Retrieved March 8, 2015, from https://www.npr.org/templates/story/story.php?storyId=16860611

Margaret J. Wheatley Quotes. (n.d.). BrainyQuote.com. Retrieved December 17, 2020, from https://www.brainyquote.com/quotes/margaret_j_wheatley_283925

Martin Luther King, Jr. quote. (n.d.). Retrieved February 9, 2015, from http://www.brainyquote.com/quotes/quotes/m/martinluth403521.html

National Geographic. (2021, May 4). *Smallpox*. Science. Retrieved April 26, 2023, from https://www.nationalgeographic.com/science/article/smallpox/

New York Post. (2013, October 20). *Little girl sold for 'organ harvesting'*. New York Post. Retrieved January 10, 2022, from https://nypost.com/2013/10/20/little-girl-sold-for-organ-harvesting/

Nsehe, M. "The Five Richest Pastors in Nigeria." Forbes. Forbes Magazine, June 24, 2017. https://www.forbes.com/sites/mfonobongnsehe/2011/06/07/the-five-richest-pastors-in-nigeria/?sh=4e45bcb56031.

Remarks by Prime Minister Hon. Dr. Kenny D. Anthony at the Formal Launch of the National Initiative to Create Employment (NICE). (2012, June 18). Retrieved January 26, 2015, from http://www.govt.lc/www/text/address_anthony_18June2012.pdf

Rincon, P. (2018, March 15). Stephen Hawking's warnings: What he predicted for the future. BBC News. Retrieved October 18, 2021, from https://www.bbc.com/news/science-environment-43408961.

Rodriguez, M., O'Hanlon, B., Vogus, A., Feeley, R., Narcisse, C., & Charles, J. (2012, January). *Saint Lucia Health Systems and Private Sector Assessment 2011*. Retrieved April 14, 2023, from https://www.hfgproject.org/wp-content/uploads/2015/02/Saint-Lucia-Health-Systems-and-Private-Sector-Assessment-2011.pdf

Russell, S. S., Gwynne, G., & Trisolini, M. (1988). (rep.). *Health Care Financing in Latin America and the Caribbean, Health Care Financing iv St. Lucia and Costs of Victoria Hospital* (Ser. 5). Stony Brook, New York: State University of New York.

Sachs, J. (2005). *The End of Poverty: Economic Possibilities for* Our Time. New York: Penguin Press.

Saint Lucia Country Strategy Paper for The Banana Industry, Agricultural Diversification & The Social Recovery of Rural Communities. (1999). Retrieved February 22, 2015, from http://www.unctad.info/upload/Infocomm/Docs/bananas/sl.pdf

Saint Lucia launches $10 million National Housing Assistance Programme. St. Lucia Times News. (2021, October 6). Retrieved November 4, 2021, from https://stluciatimes.com/saint-lucia-launches-10-million-national-housing-assistance-programme/.

Schmid, J. P. (2018, December 5). *Development challenges: Is the Caribbean losing out?* Caribbean Development Trends. Retrieved April 13, 2023, from https://blogs.iadb.org/caribbean-dev-trends/en/blog-on-development-challenges-the-caribbean-is-losing-out/

Sealy-Burke, J. (2006). *Protecting children affected by AIDS in the Caribbean: Recommendations for legal reform in saint Lucia.* World Bank. Retrieved April 22, 2023, from https://documents1.worldbank.org/curated/en/287471468340831380/pdf/372060LC0Legal1VC0July200601PUBLIC1.pdf

Shahidsaless, R. R., Kuntchev, V., & Saurav, A. (2018, July). *Comparing the impact of all-inclusive and non-all-inclusive tourism models on the quality of jobs for women.* Retrieved May 6, 2023, from https://documents.worldbank.org/en/publication/documents-reports/documentdetail/391771549915452215/comparing-the-impact-of-all-inclusive-and-non-all-inclusive-tourism-models-on-the-quality-of-jobs-for-women

Shift work and health. Institute for Work & Health. (2010, April). Retrieved April 13, 2023, from https://www.iwh.on.ca/summaries/issue-briefing/shift-work-and-health

Shorrocks, A., Davies, J., & Lluberas, R. (2022). *Global Wealth Report 2022.* Credit Suisse Research Institute. Retrieved May 7, 2023, from https://www.credit-suisse.com/about-us/en/reports-research/global-wealth-report.html

Sightings, T. (2013, January 15). *How Baby Boomers Will Change the Economy. Retrieved November 15, 2020, from https://money.usnews.com/money/blogs/on-retirement/2013/01/15/how-baby-boomers-will-change-the-economy*

Simon, G. (2018, June 12). Forensic investigation concludes in Coolie Town Deaths. https://publicservice.govt.lc/news/forensic-investigation-concludes-in-coolie-town-deaths

Smikle, P. (1996, October 18). ST. LUCIA-TRADE: Uncertainty Hits Banana Industry. Retrieved September 7, 2021, from http://www.ipsnews.net/1996/10/st-lucia-trade-uncertainty-hits-banana-industry/

South African Church Attack: Five Dead after 'Hostage Situation. BBC News. BBC, July 11, 2020. https://www.bbc.com/news/world-africa-53375607.

The Fairtrade Foundation. (2009, February). *Unpeeling the banana trade.* Retrieved April 13, 2023, from https://www.bananalink.org.uk/wp-content/uploads/2019/10/unpeeling_the_banana_trade.pdf

The London School of Economics and political Science. (2015). Arthur Lewis Exhibition. Retrieved November 14, 2020, from https://blogs.lse.ac.uk/lsehistory/files/2015/01/Arthur-Lewis-exhibition.pdf

Tourism Concern. (2013). *The impacts of all-inclusive hotels on working conditions and labour rights in Barbados, Kenya & tenerife.* Retrieved April 13, 2023, from https://www.humanrights-in-tourism.net/publication/impacts-all-inclusive-hotels-onworking-conditions-and-labour-rights-barbados-kenya

Trafficking in persons. United Nations: Office on Drugs and Crime. (2023, January). https://www.unodc.org/unodc/data-and-analysis/glotip.html

Trilling, D. (2019, September 13). *'Protecting the European Way of Life' from migrants is a gift to the far right."* The Guardian. Retrieved October 26, 2020, from https://www.theguardian.com/commentisfree/2019/sep/13/protecting-europe-migrants-far-right-eu-nationalism

Tuthill, K. (2014, April 23). *Teacher Facing Child Porn Charges Held Without Bail.* WCVB. http://www.wcvb.com/news/cambridge-middle-school-teacher-charged-with-possessing-child-pornography/25537506#ixzz2zHc3JObk

UNICEF Office for the Eastern Caribbean Area. (2021). (rep.). *Generation Unlimited: The Well-being of Young People in Saint Lucia.* Christ church, Barbados.

United Nations, Food and Agriculture Organization. (2013). (Issue brief). *Praedial Larceny in the Caribbean* (Ser. 3).

United Nations. (2009). (rep.). *The Millennium Development Goal Report 2009*. NY.

United Nations. (2012). (rep.). *The Millennium Development Goal Report 2012*. NY.

United Nations. (2013, September 5). *Secretary-General's remarks at a G20 working dinner on "sustainable development for all" [delayed in transmission]*. Retrieved April 27, 2023, from https://www.un.org/sg/en/content/sg/statement/2013-09-05/secretary-generals-remarks-g20-working-dinner-sustainable

United Nations. (2014). (rep.). *The Millennium Development Goal Report 2014*. NY.

United Nations. (2015). (rep.). *The Millennium Development Goal Report 2015*. NY.

United Nations. (2021). (rep.). *The Sustainable Development Goal Report 2021*. NY.

United Nations. (2021). *Financing for Sustainable Development Report 2021: Inter-agency Task Force.*

United Nations. (n.d.). *The Sustainable Development Agenda - United Nations Sustainable Development.* United Nations. Retrieved April 9, 2023, from https://www.un.org/sustainabledevelopment/development-agenda/

United Nations. Department of Economic and Social Affairs. Population Division. *(2019, June). World Population Prospects 2019: Highlights. Retrieved November 14, 2020, from https://population.un.org/wpp/Publications/Files/WPP2019_10KeyFindings.pdf*

United States Congress. U.S. House of Representatives, Committee on the Judiciary. (2019, March 27). VIOLENCE AGAINST WOMEN REAUTHORIZATION ACT OF 2019. https://www.congress.gov/116/crpt/hrpt21/CRPT-116hrpt21.pdf

United States Department of State. (2006, June). *Trafficking in Person Report.* Retrieved May 8, 2023, from https://2009-2017.state.gov/documents/organization/66086.pdf

United States Department of State. (2013). *Trafficking in Persons Report 2013*. Retrieved February 9, 2015, from http://www.state.gov/j/tip/rls/tiprpt/2013/index.htm

United States Department of State. (2022). Saint Lucia 2022 human rights report. https://www.state.gov/wp-content/uploads/2023/02/415610_SAINT-LUCIA-2022-HUMAN-RIGHTS-REPORT.pdf

US child sex images ring broken. (2014, March 18). Retrieved February 18, 2015, from http://www.bbc.com/news/world-us-canada-26639092

Walker, J., Pearce, C., Boe, K., & Lawson, M. (2019). (rep.). *The power of education to fight inequality*. Oxfam International.

World Bank Group. (2015, September 30). *FAQs: Global Poverty Line Update.* World Bank. Retrieved April 26, 2023, from https://www.worldbank.org/en/topic/poverty/brief/global-poverty-line-faq

World Bank. (2018, July 21). *Saint Lucia Health System Strengthening Project (P166783).* Retrieved April 22, 2023, from https://documents1.worldbank.org/curated/en/670041531772399496/pdf/Project-Information-Document-Integrated-Safeguards-Data-Sheet-Saint-Lucia-Health-System-Strengthening-Project-P166783.pdf

World Health Organization. (n.d.). *WHO coronavirus (COVID-19) dashboard.* World Health Organization. Retrieved September 4, 2022, from https://covid19.who.int/.

About the Author

Barthelemy Charles is a citizen of St. Lucia, a small island developing state located in the Caribbean. He grew up in a rural community where farming, particularly banana and sugarcane cultivation, was the main source of employment. He worked in the banana industry in various capacities for a period of about twenty years. It was primarily his interactions with banana farmers and the struggles they faced that sparked his interest in human development and social justice.

He received his academic training in international development and human services. He has acquired additional knowledge and skills working with children who are experiencing homelessness and other disadvantaged populations. To highlight issues of social importance, he published some of his research and opinion pieces as a freelance writer/contributor in the two main newspapers on the island of St. Lucia. He is also a musician and wishes to use audio/video production for public sensitization and social development.

Made in the USA
Monee, IL
04 November 2026

67378104R00066